The Animal World

The Indian leopard *(Panthera pardus fusca)* — one of the largest, most endangered cats

The World Book
Encyclopedia of Science

The Animal World

World Book, Inc.

a Scott Fetzer company

Chicago

ACKNOWLEDGMENTS

Consultant Editors
Mike Janson and Joyce Pope

Consultants and Contributors
Ray Aldridge   Nicole Bechirian   Robert
Burton   Jonathan Elphick   Thom Henvey
Casey Horton   Dorothy Jackson   Mike
Janson   Jinny Johnson   Nick Law   Jane
Mainwaring   Joyce Pope   Nora Spears
John Stidworthy   Jude Welton   Peter White

Artists and Designers
Eric Drewery   Mick Gillah   Nicki Kemball
Aziz Khan   David Parker   Mick Saunders
Charlotte Styles   Alan Suttie

Bull Publishing Consultants Ltd
Wendy Allen   Harold Bull   John Clark
Eric Drewery   Kate Duffy   Ursula Fifield
Nicola Okell   Martyn Page   Polly Powell
Hal Robinson   Sandy Shepherd

Color Separation by
SKU Reproduktionen GmbH & Co. KG,
Munich

Setting
Typoservice Strothoff GmbH, Rietberg

© Verlagsgruppe Bertelsmann International,
GmbH, Munich 1984

Published by World Book, Inc.,
Chicago, 1985; revised edition, 1987

ISBN 0-7166-3192-X
Library of Congress No. 86-50622

Printed in the United States of America

F/HG

CONTENTS

PREFACE

The quest to explore the known world and to describe its creation and subsequent development is nearly as old as mankind. In the Western world, the best know creation story comes from the book of Genesis. It tells how God created the Earth and all living things. Modern religious thinkers interpret the Biblical story of creation in various ways. Some believe that creation occurred exactly as Genesis describes it. Others think that God's method of creation is revealed through scientific investigation. *The Animal World* presents an exciting picture of what scientists have learned about the enormous variety of the Earth's animal life.

The editorial approach
The object of the series is to explain for an average family readership the many aspects of science that are fascinating and vitally important for an understanding of the world today. The books have therefore been made straightforward, concise and accurate, and are clearly and attractively presented. They are also a readily accessible source of scientific information.

The often forbidding appearance of traditional science publications has been avoided. Approximately equal proportions of illustrations and text make the most unfamiliar subjects interesting and attractive. Even more important, all the drawings have been created specially to complement the text, each explaining a topic that can be difficult to understand through the printed word alone.

The thorough application of these principles has created a publication that encapsulates its subject in a stimulating way, and that will prove to be an invaluable work of reference and education for many years to come.

The advance of science
One of the most exciting and challenging aspects of science is that its frontiers are constantly being revised and extended, and new developments are occurring all the time. Its advance depends largely on observation, experimentation and debate, which generate theories that have to be tested and even then stand only until they are replaced by better concepts. For this reason it is difficult for any science publication to be completely comprehensive. It is possible, however, to provide a thorough foundation that ensures any such advances can be comprehended — and it is the purpose of each book in this series to create such a foundation, by providing all the basic knowledge in the particular area of science it describes.

How to use this book
This book can be used in two basic ways.

The first, and more conventional, way is to start at the beginning and to read through to the end, which gives a coherent and thorough picture of the subject and opens a resource of basic information that can be returned to for re-reading and reference.

The second allows the book to be used as a library of information presented subject by subject, which the reader can consult piece by piece, as required.

All articles are presented so that the subject is equally accessible by either method. Topics are arranged in a logical sequence, outlined in the contents list. The index allows access to more specific points.

Within an article scientific terms are explained in the main text where an understanding of them is central to the understanding of the subject as a whole. Fact entries giving technical, mathematical or biographical details are included, where appropriate, at the end of the article to which they relate. There is also an alphabetical glossary of terms at the end of the book, so that the reader's memory can be refreshed and so that the book can be used for quick reference whenever necessary.

All articles are relatively short, but none has been condensed artificially. Most articles occupy two pages, but some are four, or occasionally six to twelve, pages long.

The sample two-page article opposite shows the important elements of this editorial plan and illustrates the way in which this organization permits maximum flexibility of use.

(A) **Article title** gives the reader an immediate reference point.

(B) **Section title** shows the part of the book in which a particular article falls.

(C) **Main text** consists of approximately 850 words of narrative information set out in a logical manner, avoiding biographical and technical details that might interrupt the story line and hamper the reader's progress.

(D) **Illustrations** include specially commissioned drawings and diagrams and carefully selected photographs which expand and clarify the main text.

(E) **Captions** explain the illustrations and connect the textual and visual elements of the article.

(F) **Annotation** of the drawings allows the reader to identify the various elements referred to in the captions.

(G) **Theme images,** where appropriate, are included in the top left-hand corner of the left-hand page, to emphasize a central element of information or to create a visual link between different but related articles. In place of a theme image, certain articles have a list of terms, units and abbreviations relevant to the article concerned.

(H) **Fact entries** are added at the foot of the last page of certain articles to give biographical details, physical laws and equations, or additional information relating to the article but not essential to an understanding of the main text itself.

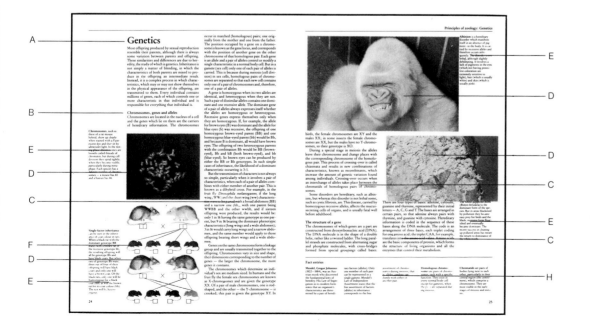

INTRODUCTION

Flight is one of the most distinctive features of birds, which separates them into their own group within the animal kingdom. There are, however, birds that no longer fly, but most still retain the feathers that aid flight in birds that do.

The twentieth century may be seen by future historians as a period in which people throughout the world learned to see the study of animals in a fresh light. The change in attitude had already begun following the publication of Charles Darwin's theory of evolution, which burst on the world in 1859. Its result was revolutionary.

Among the ancient Greeks, Aristotle had pointed the way for modern zoology, in his studies of the animal kingdom. Yet centuries later, in the Middle Ages, animals were still regarded as things that had been created simply for the use and entertainment of mankind. Man was then regarded as the lord of creation. Even the word

"zoology" was unknown until the middle of the sixteenth century. From then until the beginning of the present century the study of animals was no more than the province of learned men with time on their hands.

Until the second decade of this century, the word "zoology" had little place in the vocabulary of the man in the street. University students who elected to spend their time studying zoology were regarded as "a little odd" and were told that there was no future in the study. But to those with foresight the first rumblings of a tremendous upheaval could be heard. How and when it happened is difficult to trace. Looking at the books and university courses available around 1920 we can compare what was happening then with the state of things today.

In 1920 if you went into a bookshop and asked for books on zoology, the assistant would take you to a dim corner and point to a shelf on which were a few somberly-bound textbooks — this would happen in the best of bookshops. In the others, to ask for books on animals was to risk having the assistant, most likely, stare at you as if you had taken leave of your senses. Today, any bookshop worthy of the name has rows of shelves laden with such books, or even a section of the shop devoted to nothing else.

The difference between the books themselves these days is no less spectacular. The old textbooks were illustrated with artists' sketches. Those that were illustrated with photographs contained black-and-white images, mainly of stuffed animals or zoo specimens. Today good books on zoology are filled, page after page, with colored photographs of live animals in the wild, and colorful informative diagrams.

Formerly, in the more enlightened schools, Nature Study might be taught in the lower classes, and this consisted mainly of botany. It was not until the 1950s that the Association of British Zoologists was formed, its main purpose being to press for the teaching of zoology in schools. Universities offered courses in zoology, but the zoology department was probably in some odd corner of the building, bearing all the marks of having been added as an afterthought.

It was, however, the courses that were taught which gave the firm clues as to the standing of the subject in academic circles. The classification of animals was a first consideration. Just as the grocer or the haberdasher had his goods arranged on neat shelves, the better to be able to find them easily, so the animal kingdom was divided into phyla, classes and families the better to pigeonhole knowledge and find one's way among the million or more known species. But the main emphasis was on comparative anatomy which involved lectures on internal organs, especially the skeletons, together with the cutting-up of dead animals, known as dissection. So strong was this slant in the total teaching of zoology that students were talking about it as necrology (the study of dead things) as contrasted with biology (the study of living animals).

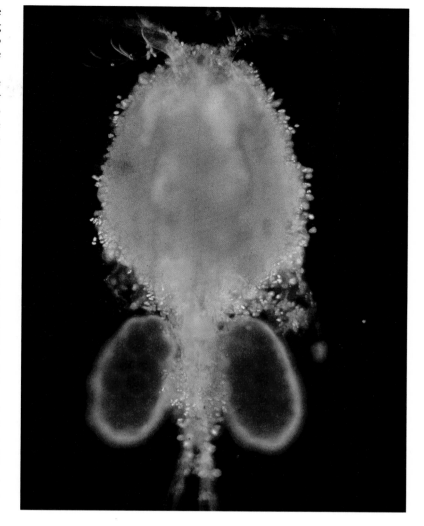

Also about the middle of the century several important events took place. Scientists began to pay more attention to the way animals behaved rather than how they were made. They began to experiment with them in the sense that they watched what they did and experimented with the causes that influenced their behavior. At first these experiments were done in the laboratory. Soon the studies were taken into the field. With the knowledge gained in laboratory studies, wild animals were watched or, as we now prefer to say, they were observed, free and unfettered, with the observer in hiding. This new science was called ethology. The name dated back to the seventeenth century and was used first to mean the study of ethics or morals, and later for the study of character.

At the same time more attention was given to the relationship between the animal and its environment, which had been known simply as natural history but later became known as ecology. These words ethology and ecology are two examples of how the language used in zoology was changing and it indicates how scientific language has changed in the last half century or so. Nowhere is this more brought out than in that word, so commonplace today — "gene." The word was coined in 1913. By the early 1920s

Animals and plants often live together in symbiosis — each feeding from the other. Green algae can be seen here living in such a relationship under the skin of a cyclops, a tiny crustacean. Despite its size this animal and many others like it play an important role in marine food chains. They feed on phytoplankton and themselves form a large part of the diet of marine animals.

African elephants wade across a lake in Zambia. They are the largest living land animals and so need to eat vast amounts of vegetation. This appetite is often the source of acute problems if their range is restricted — by human settlements, for example — and the vegetation around them is quickly consumed. They are a great worry to conservationists, who see the need for restricting elephants to game reserves to protect them, but also acknowledge that these animals need far more space than is left for them.

university lecturers, who glibly used the word "chromosome," were telling students that there were parts of the chromosome which determined the course of heredity. They did not know what these parts were but knew that they were microscopic and for want of a better name called these microscopic particles "ids."

Before long the word "gene" had crept into zoological language, replacing "id." Such was the speed at which zoology and other biological sciences were advancing, that a whole new science (genetics) was beginning to predominate. In the short space of time from 1913 genetics has developed an importance and weight which, with genetic engineering, may in the near future be fundamental to the welfare of the whole of creation — plant, animal and human.

Another important development was the invention of the electron microscope toward the end of the 1920s. The first microscopes enabled scientists to study animal structures magnified to ten times their actual size. But soon this was improved to a hundred times. By the time zoologists were talking about "ids" microscopes were available that allowed the investigator to see things a thousand times their life-size. The electron microscope now magnifies a hundred thousand times. The zoologist fortunate enough to have access to this fabulous new invention was able to see face-to-face the tiny particles of living matter, the genes that meant so much to our understanding of life itself.

At the other end of the scale from microscopic studies a new activity in macroscopic studies burst forth in zoology, represented by phenomena such as butterfly-collecting and bird-watching. They were to become important social activities, as well as constituting a significant trend in zoology.

Toward the end of the nineteenth century butterfly-collecting had held sway. During that period a naturalist was always portrayed in cartoons as a bearded gentleman in a tweed suit with knickerbockers brandishing a net on a long handle, chasing a butterfly. Butterflies were collected because of their bright colors. They were mercifully killed, so as not to damage them, and stuck with pins in drawers in glass-fronted cabinets, making a brave show that pleased the eye. These collections led, of necessity, to studying the habits of these insects, to know where and when to find them. Their classification became essential, and eventually this seemingly harmless pursuit led to a vast expansion of entomology (as the study of insects was called), itself an offshoot of zoology.

People have always been interested in birds, partly because of their songs, but more especially for their colors. But in the late 1920s and early 1930s, with the advent of the automobile as a popular means of transport and a universal impulse to get out into the country, bird-watching gained popularity in a most surprising manner. The growth of the hobby was truly phenomenal, as were its side effects. The class Aves (birds) soon became the best-documented group of animals, with the most stable classification. Not the least of the subsidiary fields of study was that of behavior. Bird-watching and its side effects could be justifiably described as a natural history mass movement with the general public, by their enthusiasm, encouragement and, not least, their insistent thirst for knowledge, unwittingly pressing the professional ornithologists to greater efforts in their researches.

With so large a proportion of the population studying wildlife for themselves came the call for conservation. From time to time during past centuries far-sighted people have realized that mankind was in danger of destroying his own environment. Theirs were, however, voices crying in the wilderness. Then in 1872 the first national park, Yellowstone, was founded in the United States (more for its scenic value than its living occupants). Thirty years later the Sable Game Reserve, later called the Kruger National Park, was established in South Africa. Since 1925, when the Albert National Park was founded in the Congo (now Zaire), the movement has spread until now the land surfaces of the world are plentifully dotted with parks and nature reserves. The movement towards conservation led to the formation of the International Union for the Conservation of Nature (IUCN).

Even the most ardent enthusiasts for the IUCN realized that they needed the support of the man-in-the-street, if its objectives were to be fully achieved. As the science of zoology came into its own, this support was forthcoming — carried along especially by the increasing popularity of bird-watching, which itself was due in no small part to modern technical achievements such as color photography, color printing and, not least, the perfection of television.

Once again, there was something resembling a mass surge. Ordinary people everywhere were becoming aware that all life is interdependent, that pollution is a destructive force to be guarded against and that wild animals in particular are in danger of disappearing from the face of the earth. It was no surprise to find that the World Wildlife Fund should have been an instant success. The world was ready for it.

A few years before the outbreak of World War II there appeared on railway stations in London, posters proclaiming that our attitude to animals conditioned the way we treated our fellow humans. It was left to those reading these posters to put their own interpretation on this message.

Knowledge brings understanding and sympathy. An increasing knowledge of zoology could bring greater understanding of and sympathy for animals and humans. Perhaps we are seeing, in the many protest movements together with the increasing interest in conservation, the beginning of an upsurge, another *levée en masse,* in tune with the spirit of that poster message. This could bring us appreciably nearer the much-talked-of brotherhood of man.

Insects are the most successful animals on earth — there are nearly a million species. The reason for their success is their tremendous adaptability which has allowed them to survive for over 400 million years through major climate changes which have caused the extinction of other animals. This ability to survive will probably ensure their success in future cataclysmic changes.

Principles of zoology

There are almost a million different species of animals living in the world today, and new ones are constantly being discovered. They range in size and complexity from microscopic single-celled creatures to the giant blue whale, weighing up to 180 tons. Faced with the task of bringing some sort of order to the living world, zoologists consider two key factors: definition (is it an animal?) and classification (what place does it occupy in the animal kingdom?).

What is an animal?

Apart from inanimate objects, we also share our world with living organisms, including plants and animals. The difference between inanimate objects and living things is not always apparent — crystals grow, for example — but only animate organisms have the fundamental characteristic of energy change known as metabolism.

Almost all the world's energy is provided initially by the sun. Plants trap a proportion of this energy and use it, through photosynthesis, to build up the sugars and starches that form a major part of their tissues. During this process they take minerals and water from the soil, and carbon dioxide from the atmosphere; they release into the air oxygen, which is a waste product of their metabolism.

Animals, however, cannot use such simple chemicals to build up the complex substances from which they are made. Instead, they must eat and digest food, and reorganize the products of digestion to form their own tissues. But in order to release the potential energy of their food, animals require oxygen, and during these metabolic processes they emit carbon dioxide as a waste product.

Thus plants and animals are complementary, each depending on and using the other's wastes which are constantly recycled. This great continuous energy flow is aided by a host of decomposers, especially bacteria and fungi. They break down dead material and waste into a form which may be used again by plants.

The distinction between a plant and an animal is often blurred among the simplest organisms, because some contain chlorophyll and can build their own food, but also hunt and feed. Of the more developed organisms, however, animals may generally be distinguished more easily by their quick responses to stimuli and their considerable powers of locomotion. In addition, only plants possess cellulose in their cell walls and contain the pigment chlorophyll.

Animal species

Plants and animals are generally classified according to a hierarchical system devised by the eighteenth-century botanist, Carolus Linnaeus. In this system, each unit is grouped with related forms into a larger taxon which then constitutes part of the next larger group, and so on. The term "species" is used as the basic unit of classification.

Scientists agree that a species can be an arbitrary category only, within which there is room for genetic diversity. They also agree that species change gradually, adapting themselves to environmental variations as successful genetic strains outbreed the less successful ones. A species is generally thought to be a population of organisms which are similar in structure and function, and capable of interbreeding freely in the wild. The words "freely" and "in the wild" are important qualifications because it is not uncommon for two different kinds of animal to be capable of producing offspring. A cross between a horse and a donkey, for example, produces a mule. In these kinds of crosses, however, the resulting hybrid is usually wholly sterile or lacks the necessary vigor to compete with other parental types. Consequently, such crosses do not contribute to the gene pool — that is, to the variety of genetic forms available.

Classification

The need for an internationally accepted classification is twofold: firstly to be able to identify a particular species throughout the world, and secondly to group related kinds into larger groups that reflect evolutionary relationships. Common,

Rotifers are named after their wheellike corona, which beats in a circular motion to achieve movement. These aquatic ani-

mals occur in abundance — about 1,500 species are known to exist — and are among the smallest multi-cellular organisms.

or vernacular, names are usually too imprecise for scientific use.

The lion, for example, is designated by the scientific name *Panthera leo*, which indicates that it is the species *leo* belonging to the genus *Panthera.* This genus also includes the jaguar *(P. onca)* and the tiger *(P. tigris).* These animals all belong to the family Felidae (the cat-like animals) which, in turn, belongs to the order of flesh-eating animals called the Carnivora. This order also includes other animals with similar features such as tooth arrangement and skull form.

As a member of the order Carnivora, the lion does not much resemble the anteater, for instance, which belongs to the different order Edentata ("toothless"), but both have mammalian features and so are designated members of the class Mammalia. All mammals have a vertebral column and are therefore placed in the subphylum Vertebrata, along with fishes, amphibians, reptiles and birds. The vertebrates share with their apparently dissimilar relatives the lancelets and sea squirts an axial stiffening rod (a notochord) and all are placed in the phylum Chordata.

Using this phylogenetic system of classification — species, genus, family, order, class, subphylum and phylum — it is possible at each level to see the relationship between animals of different species. But only the genus and species names are required to uniquely identify an animal. Thus the binomial classification of the lion is *Panthera leo.*

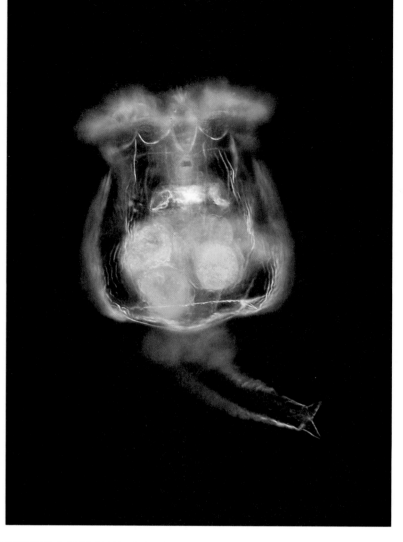

Aquatic environments are rich in microscopic organisms which provide a good food source for higher animals. Lakes such as this, in tropical Africa, are frequently inhabited by the Greater and Lesser Flamingo, which cohabit harmoniously because of the variety of food available. The Greater Flamingo sifts worms, mollusks and small arthropods from the mud, whereas the Lesser Flamingo filters blue-green algae only from the lake bottom. These birds exist easily in this environment because their long legs and necks facilitate wading and feeding in deep water.

The animal cell

The term "cell" was first used by the English naturalist Robert Hooke in 1665, to describe the "great many little boxes" he saw when viewing a thin slice of cork through a microscope. The word was derived from the Latin word for a small room (*cella*). Today, with the aid of electron microscopes, we have a more detailed view of living plant and animal cells, but Hooke's observation was the first time that living matter was recognized to be built of basic units rather than of continuous material.

Some of the cells of which plants and animals are composed have a particular function and are therefore not all exactly the same; even so they have some features in common. An aggregation of like cells forms a tissue — for example, nerve cells make up nerve tissue — and combinations of tissues in turn form an organ, such as the brain.

The structure of cells

A cell consists of a nucleus, which is surrounded by a jellylike substance called the cytoplasm, enclosed in a cell membrane. This cell membrane (also called the plasma membrane) controls the passage of substances into and out of the cell. The membrane is only 0.00001 millimeter thick and is made up of layers of lipid (fat) molecules sandwiched between layers of protein. Because the cell membrane allows only certain chemicals to move in and out of the cell but prevents the passage of others, it is said to be semipermeable.

The cytoplasm contains several structures which are classified into two basic types — inclusions and organelles. Inclusions are substances that are stored temporarily in the cell, and include fat globules and excretory products. The orga-

nelles are permanent structures and each has a vital role in maintaining the continuing life of the cell.

Organelles

Most of the cytoplasm of a mature cell is filled with elaborate folded membrane systems called endoplasmic reticulum. One type is rough and is involved with the synthesis of proteins; the other is smooth and is concerned with the manufacture of fat molecules. On one side of the membranes lie the ground substance of the cytoplasm and the soluble proteins of the cell; on the other there are many enclosed pockets called cisternae. It seems likely that the endoplasmic structures play an important role in the transport of substances throughout the cell. They also provide a large surface area where essential chemical reactions take place; many enzymes which are important for the cell's metabolism are found on the cytoplasmic side whereas their products are found in the cisternae.

The granular structures that are attached to the walls of the rough endoplasmic membranes are called ribosomes. But not all of the cell's ribosomes are attached — some swim freely in the cytoplasm. Those that are attached to the endoplasmic reticulum are concerned with the building of proteins to be transmitted to other parts of the body; those that float freely, however, aid the synthesis of the proteins which remain in the cell.

Structures similar to the smooth endoplasmic reticulum are found near the cell nucleus, and are called Golgi bodies. In fact, the smooth reticulum seems to be continuous with these bodies. The smooth membranes of these structures enclose large flattened cisternae. The role of Golgi bodies appears to be the collection and storage of the protein substances that are produced from the cis-

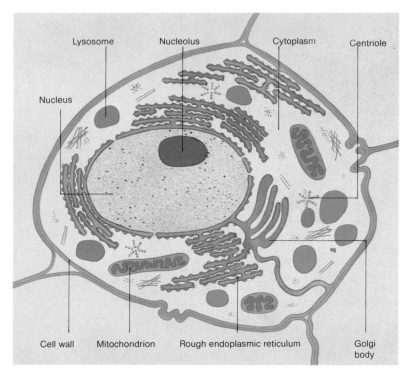

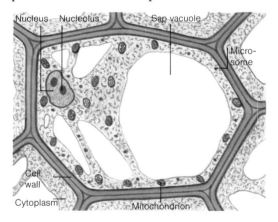

An animal cell (left) is characterized by the various structures within it, chief of which are the nucleus (with its nucleolus), mitochondria and folded layers of endoplasmic reticulum, with attached to them the protein-building ribosomes.

A typical plant cell (above) differs from an animal cell in that it has rigid cell walls (the spaces within which are filled with cellulose); a large permanent vacuole in the centre of the cytoplasm; and chlorophyll, located in the chloroplasts.

ternae of the rough endoplasmic reticulum. Once they are collected, the individual packages of protein, which are surrounded by a membrane, are transported to the cell for secretion.

The rough endoplasmic reticulum also produces packages of enzymes called lysosomes. Under acid conditions these bodies rupture and release their enzymes which break down the major components of the cell. If the cell is damaged in any way, the lysosomes release their enzymes which destroy the cell itself. Their use lies in the destruction of damaged cells or of those that are no longer needed, and in the digestion of "foreign" cells such as bacteria.

Mitochondria are another group of organelles found in the cytoplasm. They are elongated and have a double membrane which encloses a fluid-filled interior. The inner membrane comprises a series of deep folds called cristae. The mitochondria are the powerhouses of the cell, because they generate the energy that is needed to keep the cell's essential processes going. The enzymes needed to extract energy are located on the inner membrane of the mitochondria, and appear to be arranged in such a way as to derive maximum efficiency from the process. They oxidize nutrients and release energy in the form of ATP (adenosine triphosphate) which is used in the syntheses of cell materials. This oxidative process is called internal or cellular respiration.

The nucleus

The nucleus controls all the cell's activities. It is spherical or elliptical and is bounded by two membranes which together form the nuclear membrane. The outer membrane seems to be an extension of the rough endoplasmic reticulum and has several small pores through which nuclear material and large molecules pass.

Inside the nuclear membrane is a substance called the nucleoplasm which contains the chromosomes and nucleolus. The nucleolus is a spherical body which is dense in ribonucleic acid (RNA) and is the active center of protein and RNA manufacture. The chromosomes are composed of deoxyribonucleic acid (DNA) and are the blueprints for the cell's structure in the form of a genetic code. Chromosomes can be seen as thickened rods when a cell is actively dividing, but in the resting phase, the DNA is distributed throughout the nucleoplasm, when only fine threads of chromatin can be seen. These threads form the basis of the chromosomes as they thicken up again prior to cell division.

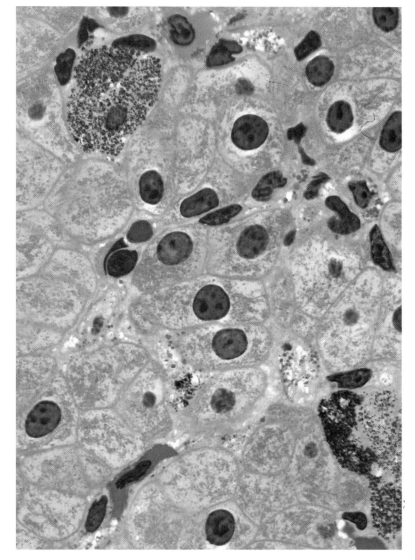

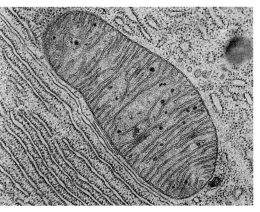

Cells are stained (above) so that they can more readily be studied. These liver cells from a salamander would otherwise be impossible to see.

All animal cells have at least one mitochondrion (left) and usually many more. These powerhouses are the sources of the energy-forming processes of the cell.

Fact entries

Energy generation in the cell starts when the products of the breakdown of carbohydrates, fats and proteins by enzymes in the process of digestion are further broken down in the cell into simpler compounds. Hydrogen atoms from them are combined with oxygen atoms in the cell to form water. The reaction between water and adenosine triphosphate (ATP) causes it to lose its third phosphate group and it becomes adenosine diphosphate (ADP), which is stored in the mitochondria. The hydrolysis — the chemical reaction of any compound with water — of the third phosphate group results in the liberation of energy which is used, for example, in muscular contraction.

Protein synthesis is the building-up of proteins which takes place in the ribosomes (particles of the cell to which proteins attach themselves). A protein is composed of amino acids, which are brought to the ribosome by transfer ribonucleic acid (RNA). The code for the arrangement of these amino acids is brought from the deoxyribonucleic acid (DNA) (which stores the genetic code for the cell) in the nucleus to the ribosome by a molecule of messenger RNA.

Anatomy and physiology

All organisms, whether plant or animal, must reproduce, grow, breathe, eat, excrete and be able to respond to their environments; special areas of animals' bodies are often organized to accomplish each of these tasks. These specialized cells may vary from animal to animal, but the chemical processes involved in the activity are usually the same. Differences in anatomy have arisen because, among other reasons, of the emergence of life from the water onto land, which has meant that various features have had to be modified to accommodate the exigencies of terrestrial life. But one of the most important developments in the pattern of anatomical organization is the occurrence of segmentation, first seen in the earthworm. It occurs in all higher groups, even if only in the embryo stage.

Structural variations

Three basic support systems exist in the animal kingdom: a water-based, or hydrostatic skeleton; a hard outer case, or exoskeleton; and a hard internal support system, or endoskeleton.

The hydrostatic skeleton is a simple fluid-filled cell and is found in animals such as amebae. In higher invertebrates such as coelenterates, nematodes, annelids and echinoderms, water held inside the body is shifted by muscles to bring about a change in shape or position of the animal. But most invertebrates have an exoskeleton which is segmented and bears jointed legs. Muscles, which enable these animals to move, are attached to the inside of the skeleton.

The main problem with this structural arrangement is that the exoskeleton weighs a lot and

The Greater Gliding Possum, *Schoinobates volans,* planes from one treetop to another on a membrane which stretches from its neck to its feet. This structural adaptation to its environment — heavily forested areas — avoids it having to run down one tree to get up another.

The supporting structures of animals consist of the hydrostatic skeleton or fluid-filled cell (A) of some protozoa and coelenterates, the exoskeleton or hard, outer covering (B), found in other invertebrates, and the endoskeleton which exists in all vertebrates, such as fishes (C) and cats (D).

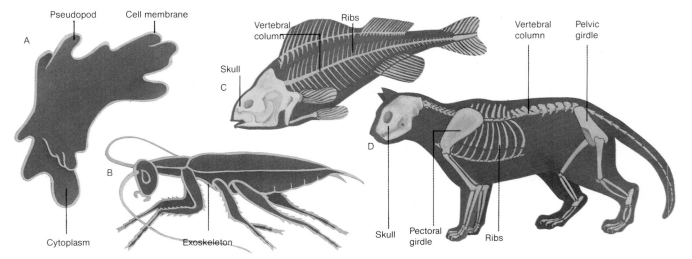

eventually becomes too heavy for the internal muscles to lift. This may be the reason for the small size of most invertebrates with such a structure.

In contrast, vertebrates have an endoskeleton which is an internal set of bony supporting structures. The backbone is strengthened by girdles to which limbs are attached. Each bone of the skeleton has muscles fixed to its surface towards the end of the bones. These muscles work in antagonistic pairs, one contracting while the other relaxes. The joints of the vertebrate body are enclosed in a synovial capsule and are lubricated by synovial fluid.

The basic structure of an animal also determines how it will grow, and the growth patterns of animals differ greatly. Considering two extremes, amebae and mammals, the first increases in size until it divides into two smaller individuals; mammals, too, grow to an adult size at which time they mature and become reproductively active, but they have special growth centers near the tips of their long bones where the cell division that enables growth takes place; in addition their skull bones are separated by cartilage which allows the skull to grow — it stops when the bones meet.

Respiration

Most animals need the chemical processes that result from gas exchange to survive. The development of respiratory organs within the animal kingdom has resulted in a variety of forms. But from the simplest to the most complex arrangement they are mainly concerned with two processes: external respiration, which involves the intake of oxygen and its transport to each cell in the body, and internal respiration which is concerned with the chemical processes that actually use the oxygen.

External respiration finishes when the oxygen crosses the respiratory surface and enters the cells. This surface, where the exchange of gases takes place, varies from animal to animal, but has certain features in common: it is moist and thin (normally one cell thick); it is permeable, allowing substances in solution to pass through it; and in

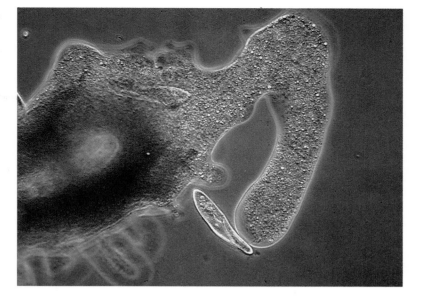

some multicellular animals, it is supplied with means of transporting oxygen to the cells, such as blood vessels. Oxygen enters single-celled animals and coelenterates through the cell wall, so the whole surface of the animal is its respiratory surface.

In the earthworm, the moist skin of the animal also acts as the respiratory surface but its body is too thick for the oxygen to simply diffuse from the outer layer to the inner cells, so a blood circulation system has developed which is able to carry dissolved oxygen to all parts of the body.

In terrestrial arthropods a series of air tubes (tracheae) run from the outer surface into the body and end in special areas of gas exchange, the alveoli. Each alveolus is kept moist so that oxygen can dissolve into the water film and thus leave the alveolus to diffuse into the body cells. Aquatic invertebrates with an exoskeleton, and aquatic vertebrates, have a gill system which acts as a respiratory surface, extracting oxygen from the incoming water.

Land vertebrates have lungs which contain the air tubes, alveoli (the respiratory surface), and a supply of blood vessels to carry the oxygen to other parts of the body. In mammals the lungs are

Single-celled protozoa, such as amebae, move by extruding their cytoplasm, which flows forward to form pseudopodia (false feet). This method of locomotion is also used to trap food. The foot flows around and envelops a food particle which is broken down by digestive enzymes.

Digestion in lower animals is simple: amebae (A) engulf food and digest it only with enzymes; insects (B) have a mouth and a simple digestive tract. But vertebrates (C and D) have a higher metabolic rate and so need to consume more; they therefore have digestive glands and long intestines.

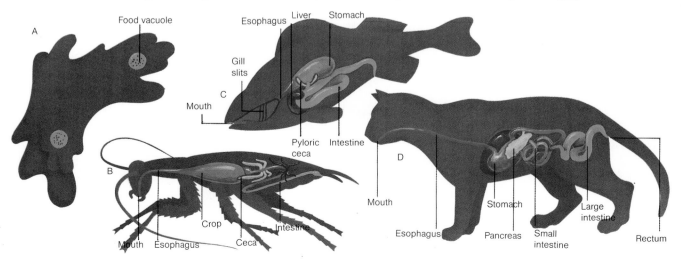

Fledgling songbirds are nestbound and unable to see at birth, and for several weeks are dependent on their parents for food. This is in contrast to many less sophisticated animals, such as larvae and young insects, which after birth are able to fend for themselves. The vulnerability of young birds makes them easy prey for predators and means that they need constantly to be guarded by their parents, from whom they learn to fly and sing.

Circulation in amebae (A) consists only of water flowing in and out of the cell through a semipermeable membrane. Insects (B) have a dorsal blood vessel which contains a 13-chambered heart. Cold, colorless blood is pumped forward by muscular contractions in the vessel and returns backward through the blood spaces to reenter the heart. Fishes (C) have a ventral blood vessel from which segmented vessels branch and run through the gills to join the dorsal vessel which distributes blood to the body. Fishes and mammals (D) have a four-chambered heart, but oxygenated blood in mammals is distributed by arteries, and veins return deoxygenated blood to the heart.

protected by a ribcage and separated from the rest of the internal organs by a muscular sheet, the diaphragm. The ribs and diaphragm act together to pump a continuous stream of fresh air into the lungs. Air entering the lung dissolves in the film of water in the alveolus chamber and so diffuses through the chamber wall into the blood vessels. The hemoglobin in the blood transports the oxygen to the cells.

The nervous system

Every animal has sensory apparatus, the increasing complexity of which can be followed from aquatic to terrestrial animals. Simple organisms like amebae are sensitive over the whole surface of their bodies to general stimuli because no particular area is responsible for collecting specific information. Jellyfish and sea anemones have identifiable nerve cells which form a nerve net across the surface of their bodies but they have little specialization and respond generally to stimuli such as light and dark, acid and alkali, hot and cold.

In worms and arthropods some nervous tissue is concentrated at the front of the body. In these animals a primitive brain joins a thick ventral nerve cord which runs the length of the body.

Smaller nerves branch from it to other parts of the body and so a basic information-collecting service is formed.

The environmental demands on vertebrates are greater than those on invertebrates. They have therefore developed a specialized complex nervous system composed of a brain and a dorsal nerve cord which is protected by the spinal column. The brain is connected to the eyes, nose, ears and other parts of the head by a series of cranial nerves. The rest of the body is served by the spinal nerves which join the spinal cord. Mammals and birds have larger brains than other vertebrates because their warm blood provides the stable internal environment necessary for the development of the brain.

The evolution of sensory organs can be seen particularly in the eyes; in some crustaceans and some worms, sight is only sensitive to light — these animals have simple clusters of photoreceptors, called ocelli, which guide them toward and away from light. In higher animals the sensory cells are so organized that an image forms inside the eye. Most vertebrates have lateral vision in which a separate two-dimensional image is formed by the eye on each side of the head, but in mammals and birds of prey the eyes have turned forward so that a single image of the same object is formed; this is stereoscopic vision and is three-dimensional.

Digestion and dentition

In all animals, whether simple or complex, food is digested by enzymes which convert it to its simplest form. Amebae simply engulf food particles and package them into a food vacuole into which digestive enzymes are secreted. These enzymes break the food down into its chemical parts. In coelenterates, food is taken in through the single opening — the mouth. The lining of the internal cavity then secretes enzymes which digest the food material, and ameboid cells inside the cavity engulf the digested food. Unused food is rejected from the body via the mouth.

Most other animals have a tube which runs from the mouth to the anus. The food enters the

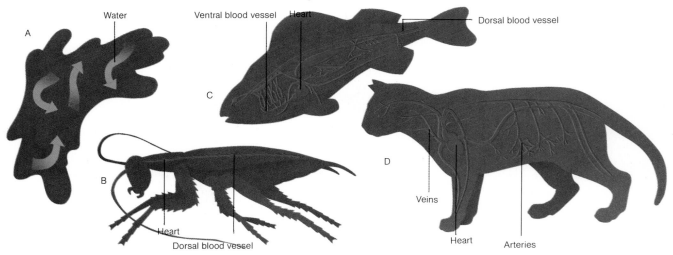

A   Water

Ventral blood vessel   Heart   Dorsal blood vessel

C

B   Heart

Dorsal blood vessel

D

Veins

Heart   Arteries

tract via the mouth which opens into the buccal cavity, where the digestive process begins. The food then passes along a muscular section of the tube, pushed by waves of contraction (peristaltic waves) until it reaches the stomach. Most of the digestive process occurs in the stomach or an area close to it. Once digestion is completed the useful materials are assimilated whereas the residue is passed out of the body via the anus.

In addition specialized glands are often present (particularly in vertebrates) which produce secretions that aid the digestive process. In mammals, for example, the liver produces bile, and the pancreas releases digestive enzymes into the small intestine. Food broken down by enzymes is assimilated into the blood supply and taken to the liver. The liver begins the elimination process of any unnecessary chemicals.

Most animals also contain microorganisms which live commensally in their digestive systems. There are also some cases of symbiotic relationships, such as some termites, which rely on protozoa in their gut to break down the wood that they eat.

Most animals merely swallow their food, making no attempt to break it up. In some groups, such as birds and some insectivores, tiny stones are present in the stomach which help to degrade the food and facilitate enzyme action. Most vertebrates use their teeth to hold food only before swallowing it. Mammals, however, chew their food. They have a lower jaw which is made of a single pair of bones with differentiated teeth. This change from other animals must be a result of the warm-bloodedness of mammals — they use their jaws much more than any other group because they need food to satisfy their high metabolic rate, and to get all the nutritional value they can from food they must start processing it in the mouth. They have developed different types of teeth for this purpose. Generally, herbivorous animals have very long jaws with a battery of flat-topped grinding cheek teeth, which break up plant material that would otherwise be impossible to digest. Meat-eating animals have sharp teeth, or canines, which are used to kill their prey. But they also have slicing cheek-teeth, or carnas-

sials, which are used to cut the meat into swallowable pieces. Omnivorous mammals have a combination of flat-topped cheek-teeth with sharp, cutting edges and spatulate incisors for biting.

Excretion

Excretion is the process that eliminates waste from the body. In many simple organisms waste diffuses out via the body wall. In larger animals, however, special organs have developed to deal with this process.

The circulatory and respiratory systems in vertebrates remove carbon dioxide and water from the body, whereas the kidneys filter out excess chemical substances brought to them by the blood. These substances, which include nitrogen from the breakdown of proteins, excess salts and sugars, hormones and water, pass out of the blood into the kidney tubule. Harmful substances then collect in the bladder and are passed out of the body in the form of urine. In some animals, such as insects and terrestrial gastropods, for whom water retention is essential, dry uric acid crystals are excreted. In others, such as fishes, for whom water loss is not as great a problem, waste is excreted as ammonia, the formation of which requires water.

The newborn young of grazing animals are fully active soon after birth — this young wildebeest could get on its feet within seven minutes of being born. They feed on mother's milk for several months but, unlike nestbound fledglings, can move around independently.

Respiration in amebae (A) involves oxygen diffusing into the cytoplasm and carbon dioxide diffusing out. In insects (B), air is conducted through tracheae to the blood, into which oxygen diffuses. Fishes (C) extract oxygen from water which passes into the mouth and out through the gills. Mammals (D) breathe with lungs, where oxygen diffuses into blood vessels.

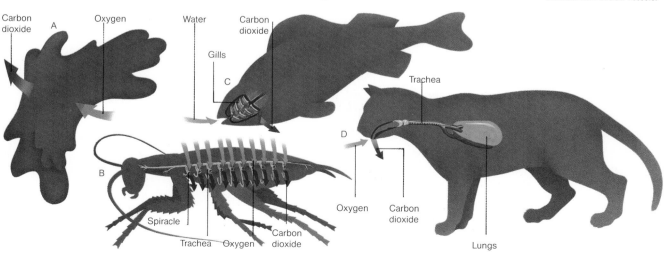

Reproduction

Death from old age, predation, disease, injury or sudden environmental change is an ever-present threat to animals. A species can survive and adapt, however, because its members reproduce themselves. When animals reach adult size they become sexually mature and able to reproduce. They do so in one of two basic ways — by asexual or sexual reproduction. Asexual reproduction involves only one parent, whereas sexual reproduction involves two adult individuals.

In many animals the development of reproductive cells is induced by hormones and pheromones. These chemical substances are activated by environmental factors such as the longer days and additional light in spring and early summer. The production of hormones also influences courtship and mating behavior. The young are usually born some time later at a time of plentiful food supply. This coincidence of birth with abundant food increases the individual's chances of survival.

Asexual reproduction

Only the more primitive animals reproduce asexually. It is the fastest means of reproduction but has the disadvantage of producing offspring that are identical to their parents. There is no shuffling of genetic material between generations, and therefore less variation within the population.

The simplest method of asexual reproduction — binary fission — is found in unicellular organisms such as amebae. These animals simply split their nucleus in two by mitosis (cell division) and the cytoplasm separates to surround each new nucleus. The parent no longer exists as a single unit but has become two "daughter cells."

Two other methods of asexual reproduction — fragmentation (or regeneration) and budding — are also found in animals which are capable of sexual reproduction. Regeneration is the process by which animals such as some annelids and jellyfish break off parts of their bodies; each part then grows into a new adult individual. Budding involves the development on the body surface of an outgrowth, or bud, which grows into a smaller copy of the parent and eventually detaches itself.

Budding is a method of asexual reproduction and is characteristic of coelenterates such as *Hydra.* It differs from regeneration in that specialized cells are involved. These cells on the body surface grow rapidly to form an outgrowth, or bud, and develop into a smaller copy of the parent.

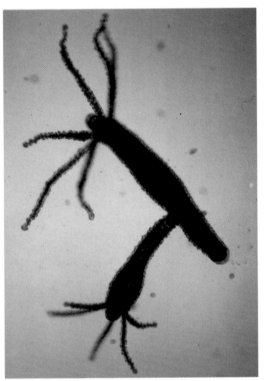

Surinam toads *(Pipa pipa)* fertilize their eggs externally. The female collects her eggs on her back, where the male covers them with sperm during amplexus. The eggs embed themselves in the female's back until they are ready to hatch.

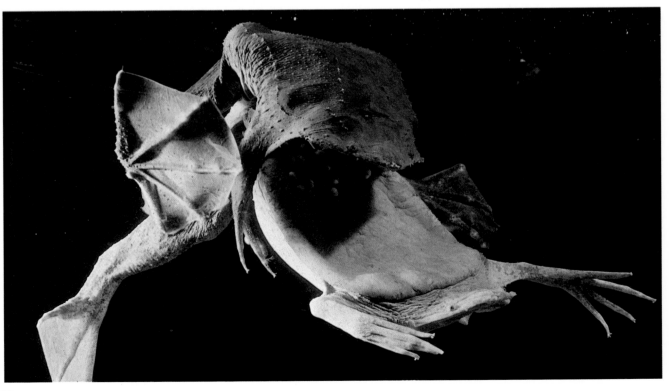

In certain corals the bud does not completely detach itself, but remains close to it attached by a thin strand of living cells. Thousands of individual animals, all derived from a single parent, may form great colonies of coral in this way.

Parthenogenesis is another form of asexual reproduction, in which an unfertilized egg grows into an organism. The individuals produced in this way are normally female, as among aphids, and can only reproduce daughters; only fertilized eggs give rise to adults of both sexes that can mate and reproduce. In some animal groups such as coelenterates, there is an alternation of sexually and asexually produced individuals, one stage being the dominant, usual state for that species, the other being a temporary phase.

Sexual reproduction

The two parent organisms involved in sexual reproduction each produce special sex cells, or gametes. Female gametes (ova) are formed in an ovary; they are usually much larger than the male gametes (in fact they are the largest single cells) and contain nutrients which feed the embryo. They are also virtually immobile. The male gametes (spermatozoa) are formed in the testes, are usually very small, and move extremely fast by means of a flagellum.

Gametes contain half the number of chromosomes that every other adult cell in the body possesses, and so are called haploid, whereas the body cells contain two corresponding sets of chromosomes (homologous chromosomes) and are diploid. This reduction in the number of chromosomes is effected by meiosis.

When the nuclei of two gametes fuse together during fertilization the resulting cell, or zygote, contains the full number of chromosomes and two sets of hereditary information, one from each parent. Not all this information is used in the new organism, but is selected according to the rules of genetics — the offspring exhibit features of both their parents but are not exact copies of either. It is this variation that confers an advantage on animals that reproduce sexually.

Sexuality and fertilization

Some animals, such as hydras, flatworms, earthworms and snails, have both ovaries and testes in each individual, and are called hermaphrodites. Tapeworms fertilize themselves, but most hermaphrodites cross-fertilize with other members of their species. In most animals, however, the male and female sex organs are found in different individuals.

Those animals that live in water reproduce there; their reproductive cells do not need to be protected because they are surrounded by an aquatic medium. But terrestrial animals do not have a watery environment in which fertilization can take place and so have developed special reproductive fluids to carry the reproductive cells. They also have special organs to effect the transfer.

For fertilization to occur it is important that

Mitosis (A) is the process by which all body cells are produced. Meiosis (B) occurs only in sex cells. They both involve several phases during which a diploid cell is split. During prophase the centrioles separate and form a spindle around the nucleolus, which disappears. The chromosomes collect at the spindle's equator with the centromeres on the spindle threads. In mitosis the centromeres divide and migrate to opposite poles but in meiosis whole chromosomes polarize. The cell then splits. In mitosis the chromosomes unwind and revert to chromatin threads. Each daughter cell contains one half of each original pair of chromosomes and replicates the other half. In meiosis the two new cells split again so that the four resulting cells are haploid, containing only one half of each original pair of chromosomes.

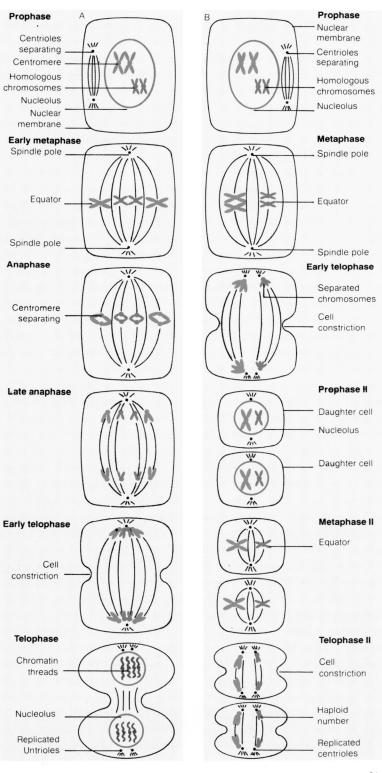

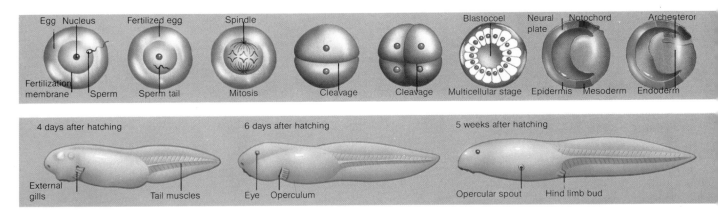

Egg Nucleus	Fertilized egg	Spindle			Blastocoel	Neural plate	Notochord	Archenteror
Fertilization membrane Sperm	Sperm tail	Mitosis	Cleavage	Cleavage	Multicellular stage	Epidermis	Mesoderm	Endoderm

4 days after hatching	6 days after hatching	5 weeks after hatching
External gills Tail muscles	Eye Operculum	Opercular spout Hind limb bud

The fertilization of a frog egg is followed by mitotic division of the egg, called cleavage. The egg cleaves into a two-, four- and then eight-celled organism. A cavity (blastocoel) forms as cleavage continues and the cells start to differentiate (gastrulation). The skin and body cells then start to develop.

Young mammals depend on their mother for food which, during the first few weeks or months of life, takes the form of milk. Parental care also includes keeping the infants clean and protecting them from predators.

the sperm and ova are deposited close to each other. The mobile sperm swim to an ovum, attracted by chemicals in the surrounding fluid, but because sperm consist mostly of genetic material, they have a very small energy store and usually cannot survive for more than 24 hours under normal conditions. For this reason most animals, especially land-dwellers, set several mechanisms in motion to ensure the meeting of sperm and ovum.

The simplest form of fertilization exhibited by higher animals is spawning, found in creatures that breed in water, such as fishes and frogs. This method requires exact timing of the release of gametes from both sexes, or the egg will not be fertilized. The female usually releases her eggs first and the male then fertilizes them by covering them with sperm.

Fertilization in aquatic animals is usually external, but land animals such as reptiles, birds and mammals rely mostly on internal fertilization. During copulation the male deposits his sperm inside the female, and the sperm swim to the ovum.

Ovipary and vivipary
After fertilization, birds and reptiles lay protected

eggs which develop outside the mother. This egg-laying process is known as ovipary. The most advantageous form of reproduction, however, is vivipary, which is exhibited by placental mammals and ensures that the fertilized egg implants inside the mother. It is nourished by the mother's bloodstream until birth, when it has reached an advanced stage of development and looks like a smaller version of the adult. Some snails, insects, fish, amphibians and reptiles are ovoviviparous, in that the embryo develops within the mother but is protected by egg-membranes until it hatches inside the mother's body.

Insects
Most insects reproduce sexually and produce yolk-filled eggs, which follow three different patterns of growth before the insect becomes an adult or imago.

The most highly evolved insects such as butterflies and bees develop by complete metamorphosis. They hatch from their eggs as larvae and do not resemble their parents. The larvae spend their lives eating and growing. This growth is normally achieved by molting. When the insect is full-grown, eating stops, and the bloated larva becomes a pupa — such as the chrysalis of a butterfly. Inside the pupal case the adult insect body develops and eventually emerges.

Another pattern of development, incomplete metamorphosis, is exhibited by the group of insects that includes dragonflies and cockroaches. The insects hatch from their eggs as nymphs, or naiads, which resemble the parents, but are wingless, smaller and sexually immature. Like larvae, nymphs eat all the time. They do not enter a pupal stage, however, but grow and shed their hard outer skin until they reach adult size when they become sexually mature.

The most primitive kinds of insects, such as the wingless bristletails (order Thysanura), hatch as miniature replicas of their parents and no metamorphosis is apparent.

Fishes and amphibians
The reproductive behavior of fishes varies considerably between species. A herring, for example, may shed tens of thousands of eggs and rely on external fertilization, whereas a dogfish reproduces by internal fertilization and produces very

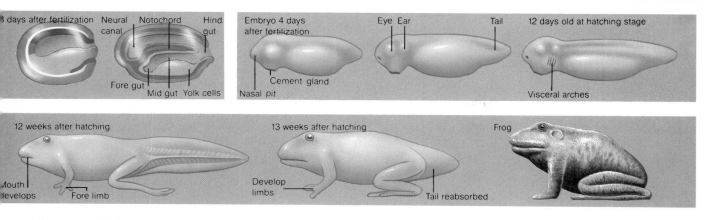

days after fertilization | Neural canal | Notochord | Hind gut

Fore gut | Mid gut | Yolk cells

Embryo 4 days after fertilization | Eye | Ear | Tail | 12 days old at hatching stage

Cement gland
Nasal pit | Visceral arches

12 weeks after hatching | 13 weeks after hatching | Frog

Mouth develops | Fore limb | Develop limbs | Tail reabsorbed

few eggs, which are enclosed in an egg purse to protect them while they develop.

Amphibians are considered to be more advanced than fishes, but they are not fully adapted to life on land and always return to water for breeding. Most amphibians fertilize externally. Frogs and toads, for example, mate after they have exchanged specific courtship signals. The mating process (amplexus) involves the male seizing the female and sitting on her, which induces both of them to release gametes into the water simultaneously. The fertilized eggs first go through a larval stage, when they hatch as tadpoles, and gradually metamorphose later into adult amphibians. In some salamanders, however, the young are born live and do not go through a larval (tadpole) stage.

Birds
Birds all reproduce by internal fertilization. Unlike amphibians they do not rely directly on water for breeding, but the hard-shelled eggs they lay each contain a "private pond" in which the embryo floats and develops. Because birds are warm-blooded the embryos must be kept at body temperature to develop. The temperature is maintained by incubation, a process which is usually carried out by the female, but which in some species is performed by the male while the female goes off in search of food. Incubation can take two to three weeks or more, before the young chicks hatch. In the case of some birds, the female continues to brood for long periods after they have hatched because the chicks are born blind, helpless, unable to feed themselves and without feathers to keep them warm.

Mammals
Mammals fertilize and develop internally. Monotremes, among mammals, are egg-layers, and have a uterus structure similar to that of reptiles. In other mammals, the uterus structure varies greatly — female rabbits and kangeroos, for example, have a double uterus and vagina, cows have a two-horned (bicornuate) uterus, and human females have a single, triangular uterus. While the embryo grows inside the mother's uterus, it is supplied with oxygen and nourishment from the mother's bloodstream via the placenta and is kept at a constant warm temperature. The mother can continue with a normal life up until a few hours before birth, after which the young mammal is entirely dependent on her for milk until it can eat solid foods.

A developing frog embryo elongates as it nears hatching and various sense plates become visible, as does a cement gland in the mouth area, and visceral arches. About four days after the tadpole emerges, tail muscles develop and external gills appear around the visceral arches. It attaches itself to plant matter with its oral sucker. About a week after hatching, an operculum grows across the gills, which wither. The eyes develop and the cement gland disappears. About six weeks after hatching a hind leg buds followed by fore legs. The mouth then widens and the tail shortens. The metamorphosis is complete at about 13 weeks.

A two-headed Pacific Gopher snake (Pituophis catenifer) is the result of the same reproductive process that creates Siamese twins. It occurs when a fertilized egg splits partially into two during an early stage of the egg's development. Mutations such as these rarely survive because their internal organs are often greatly deformed; they are therefore weaker and more vulnerable.

Genetics

Most offspring produced by sexual reproduction resemble their parents, although there is always some variation between parents and offspring. These similarities and differences are due to heredity, the study of which is genetics. Inheritance is not simply a matter of blending, in which the characteristics of both parents are mixed to produce in the offspring an intermediate result. Instead, it is a complex process in which characteristics, which may or may not show themselves in the physical appearance of the offspring, are transmitted to them. Every individual contains millions of genes, each of which controls one or more characteristics in that individual and is responsible for everything that individual is.

Chromosomes, genes and alleles

Chromosomes are located in the nucleus of a cell and the genes which lie on them are the carriers of hereditary information. The chromosomes occur in matched (homologous) pairs; one originally from the mother and one from the father. The position occupied by a gene on a chromosome is known as the gene locus, and corresponds with the position of another gene on the other chromosome of that homologous pair. Each gene is an allele and a pair of alleles control or modify a single characteristic in a normal body cell. But in a gamete (sex cell) only one of each pair of alleles is carried. This is because during meiosis (cell division) in sex cells, homologous pairs of chromosomes are separated so that each new cell contains only one of a pair of chromosomes and, therefore, one of a pair of alleles.

A gene is homozygous when its two alleles are identical, and heterozygous when they are not. Such a pair of dissimilar alleles contains one dominant and one recessive allele. The dominant gene of a pair of alleles always expresses itself whether the alleles are homozygous or heterozygous. Recessive genes express themselves only when they are homozygous. If, for example, the allele for brown eyes (B) was dominant and the allele for blue eyes (b) was recessive, the offspring of one homozygous brown-eyed parent (BB) and one homozygous blue-eyed parent (bb) would be Bb, and because B is dominant, all would have brown eyes. The offspring of two heterozygous parents with the combination Bb would be BB (brown-eyed), Bb and bB (both brown-eyed), and bb (blue-eyed). So brown eyes can be produced by either the BB or Bb genotypes. In such simple cases of inheritance, the likelihood of a dominant characteristic occurring is 3:1.

But the transmission of characters is not always so simple, particularly when it involves a pair of characteristics, when each of a pair of alleles combines with either member of another pair. This is known as a dihybrid cross. For example, in the fruit fly *Drosophila melanogaster*, if the long wing (WW) and the short wing (ww) characteristics were to be paired with a broad abdomen (BB) and a narrow one (bb), with one parent being WWBB and the other wwbb, and if sixteen offspring were produced, the results would be: only 1 in 16 having the same genotype as one parent, but 9 in 16 bearing the dominant phenotypic characteristics (long wings and a wide abdomen); 3 in 16 would carry long wings and a narrow abdomen, and the same number would apply to those offspring bearing short wings and a wide abdomen.

Genes on the same chromosome form a linkage group and are usually transmitted together to the offspring. Chromosomes vary in size and shape, their dimensions corresponding to the number of genes — the larger the chromosome, the more genes it contains.

The chromosomes which determine an individual's sex are medium-sized. In humans and the fruit fly the female sex chromosomes are known as X chromosomes and are given the genotype XX. Of a pair of male chromosomes, one is rod-shaped, and the other — the Y chromosome — is crooked; this pair is given the genotype XY. In

Chromosomes, such as these of a rat-mouse hybrid, show up clearly when stained with a fluorescent dye and then lit by ultraviolet light. In the resting phase chromosomes are loosely coiled threads of chromatin, but during cell division they spiral tightly, when they become visible, particularly during metaphase. Each species has a definite number of chromosomes — a mouse has 40 and a human has 46.

Single-factor inheritance can be seen in the inheritance of coat color in rats. When a black rat with the dominant genotype BB mates with a brown rat of the recessive genotype bb, the resulting offspring will all be genotype Bb and have black coats. But when rats of genotype Bb mate, three out of four of their offspring will have black coats and only one will have a brown coat. Of the black rats, only one will be homozygous for a black coat (BB) as will the brown rat for its coat color (bb). The rest will be heterozygous.

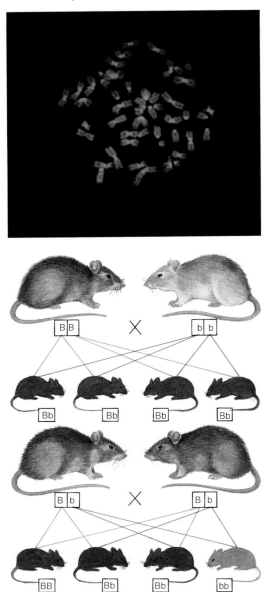

24

Albinism is a hereditary disorder which manifests itself in an absence of pigments in the body. It is carried by recessive alleles and therefore occurs infrequently. The disorder is not lethal, although slightly debilitating. It involves a lack of pigments in the eyes (which not having protective coloration are extremely sensitive to light), hair (which is usually white) and skin (which is usually pink).

birds, the female chromosomes are XY and the males XX; in some insects the female chromosomes are XX, but the males have no Y chromosomes, so their genotype is XO.

During a special stage in meiosis the alleles leave their chromosome and change places with the corresponding chromosome of the homologous pair. This process of crossing-over is called chiasmata and results in new combinations of characteristics, known as recombinants, which increase the amount of genetic variation found among individuals. Crossing-over occurs when an interchange of alleles takes place between the chromatids of homologous pairs of chromosomes.

Some disorders are hereditary, such as albinism, but whereas this disorder is not lethal some, such as cystic fibrosis, are. This disease, carried by homozygous recessive alleles, affects the mucus-secreting cells of organs, and is usually fatal before adulthood.

The structure of a gene
The chromosomes of which genes are a part are constructed from deoxyribonucleic acid (DNA). The DNA molecule is in the shape of a double helix, rather like a twisted ladder. The long parallel strands are constructed from alternating sugar and phosphate molecules, with cross-bridges formed from special groupings called bases.

There are four types of bases: adenine, cytosine, guanine and thymine, represented by their initial letters — A, C, G and T. The bases are arranged in certain pairs, so that adenine always pairs with thymine, and guanine with cytosine. Hereditary information is coded in the sequence of these bases along the DNA molecule. The code is an arrangement of three bases, each triplet coding for one amino acid; the triplet CAA, for example, translates into the amino acid valine. Amino acids are the basic components of protein, which forms the structure of living organisms and all the enzymes that control their metabolism.

The white peppered moth (Biston betularia) is the dominant form of the species. But in areas blackened by pollution they became easy prey for birds and the black, recessive form flourished and eventually became dominant. The recent success in cleaning up polluted areas has meant the return to dominance of the white moth.

Fact entries

Mendel, Gregor Johann (1822—1884), was an Austrian monk who, largely through his study of garden peas and other plants discovered the fundamental laws of heredity. His Law of Segregation in its modern form states that an organism's characteristics are determined by a pair of hereditary factors (alleles). Only one member of each pair can be represented in a single gamete. Mendel's Law of Independent Assortment states that the free assortment of factors (alleles) in inheritance corresponds to the free assortment of chromosomes during meiosis; that is, alleles combine randomly with either of another pair. His achievement was unrecognized in his lifetime.

Homologous chromosomes are pairs of chromosomes, each of which has a specific function. They exist in every normal body cell except for gametes, in which the pairs have been separated during meiosis.

Chromatids are pairs of bodies lying next to each other, particularly in their central region (the centromere), which comprise a chromosome. They are most visible in the early stages of meiosis and mitosis.

Behavior

In order to survive, an animal must be able to adapt itself to changing factors in its environment: in other words, it must respond. Its responses include simple reflexes based on instinct and learning, as well as more complex patterned behavior such as the division of labor among some communities. Mating and feeding are the most important activities and are often the motive for many rituals of behavior.

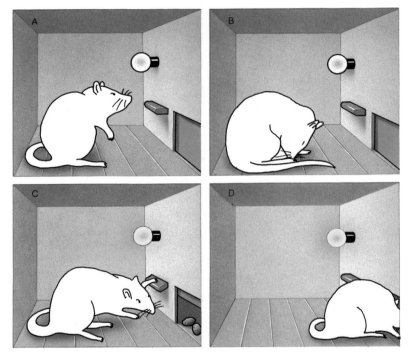

A rat placed in a Skinner box (A) will quickly learn how its surroundings work. It sits and waits (B) for the stimulus — the flashing light — when, it has learned, if it presses the lever (C) food can be obtained (D).

Instinct and learning

All animals are born with a range of innate behavior patterns. The knowledge to do certain things is controlled genetically and is called instinct. This form of inherited behavior helps an animal to fit into and survive in its environment.

In addition to instinct, animals are able to learn ways of coping with their environment and can thus modify their behavior to deal with problems that they have encountered before. The ability to learn is taken by some behaviorists to be a measure of an animal's intelligence.

But most animals use a combination of instinct and learning to adapt to their environment; the longer an animal's life span and the more complex its life style, the more learning has a role to play. A large predatory mammal, for example, has an inbuilt hunting ability which is improved by learning from the other members of its group. A fly, by comparison, which lives for only a short time, has as its sole purpose to feed and breed successfully before it dies. It does not have the time to learn how to find a mate, or where it would be best to lay its eggs. Instinct provides the fly with a fixed set of adaptive responses to its environment so that it can cope even though it has no previous knowledge of a situation. Instinct is also important when there is no parental care, when the animal is not able to learn by example.

Instinctive reaction is characterized by two features: it is the same for all members of a species, and it can be initiated by a simple stimulus. But the division between instinct and learning is not immediately obvious. All animals of a particular species show the same reaction to a particular stimulus, even if they have been isolated from

The courtship ritual of gannets *(Sula serrator),* which breed in large communities, is essential to the male in the first breeding year to enable him to establish the sex of any apparently unmated bird, because there are few superficial distinguishing sexual characteristics. In the ceremony two birds elongate and twist their necks. In subsequent breeding seasons the partners will recognize each other and come together again.

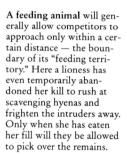

Ticks are drawn toward sunlight (A), by the smell of butyric acid, given off by warm-blooded animals (B), and the warmth of flesh into which they sink their proboscis and suck up blood (C). But experiments with a warm, water-filled balloon onto which butyric acid is smeared show that a tick is just as likely to drop onto the balloon and suck up the water.

birth, although animals also often carry out learned activities in a similar stereotyped fashion which may appear to be instinctive. Responses learned in isolation vary, however, among individuals of a species depending on how they were taught.

An animal's ability to learn and remember is, to a certain extent, dependent upon the life it leads. A worker bee, for example, must be able to learn the position of its hive and of flowers that are good for foraging, but a fly does not need this ability. It is not surprising then that experiments have proved that whereas bees can easily remember the position of plates of food, flies are less able to do so.

Insight is probably the highest form of learning — when the answer to a problem arrives in a sudden flash. The term is used to describe the rapid solution of a problem, too fast for a trial and error process. An example of this is the immediate construction by some chimpanzees of a "ladder" from objects in order to retrieve fruit that is too high to get at by stretching.

Another type of learning is imprinting, which is thought to occur only over a short, critical period when the animal is very young, or at crucial times of an animal's life, such as at the production of offspring. Imprinting occurs in both mammals and birds. Young songbirds, for example, are not able to sing an adult version of the species song if they have never heard it. They will sing, instead, a simple version of the song, without any characteristic trills. But a young bird which has heard the adult song once only will be able to sing it perfectly at a later stage because it has imprinted the song.

Small animals and birds also imprint any object that is near them soon after birth. This "maternal" imprinting ensures that the young follow the mother, as waterfowl do. But young birds can be persuaded to imprint balloons, cardboard boxes, and even human beings.

Apart from maternal imprinting, there is also sexual imprinting. An animal which has, for some reason, managed to imprint the wrong species, or an inanimate object, will not mate with its own kind. This imprinting was illustrated in an experiment where a male Zebra finch (*Taeniopygia guttata*) was raised by a pair of Bengalese finches (*Lonchura striata*). When it reached maturity it was placed in a cage with a female Zebra finch and a female Bengalese finch; the male ignored the

female of his own species and courted the Bengalese female, despite her lack of interest.

Imprinting is also thought to be olfactory and may account for a mother's recognition of her offspring, as in ungulates such as cows or wildebeest. When a mother gives birth, she imprints the smell of her offspring and is then able to identify it in a herd.

The role of sense organs
In addition to instinct, learning is closely associated with responses caused by stimulated sense organs. Most animals are aware of and respond to changes in light, gravity, temperature and noise but do so in different ways.

Many animals forage for food which they bring back to their home but they need to orient themselves in the environment and recognize signs to find their way back. Digger-wasps (family Scoliidae), for example, recognize landmarks around their burrows, but only the pattern of them — if the pattern were changed, for example, they would not immediately recognize the position of their burrows.

Stimuli such as those used in courtship displays are often spread over a few hours or days and have a cumulative effect on an animal. The effect breaks down defensive barriers between the male and female so that copulation can take place, and also brings the female into a sexually receptive state.

A feeding animal will generally allow competitors to approach only within a certain distance — the boundary of its "feeding territory." Here a lioness has even temporarily abandoned her kill to rush at scavenging hyenas and frighten the intruders away. Only when she has eaten her fill will they be allowed to pick over the remains.

Combat among the males of a herd is often the means of asserting rights over territory, females and the dominant position in the herd. Among animals such as Thomson's gazelle *(Gazella thomsoni)*, males and females form separate herds; a few dominant bucks lead the bachelor herds, and during the breeding season try to attract females into their territory to mate with them.

This is necessary — indeed it can be vital — particularly in scorpions and spiders which at the approach of an animate object, even if it is a member of its own species, assume an aggressive position. In such cases if a male intending to mate does not approach the female cautiously he may be attacked or killed.

Animals respond to stimuli that are quite different from those discernible by human beings. For instance, it was a puzzle for many years to zoologists why some flowers that depended on bees for pollination were completely white and apparently had no guiding pattern (nectarguides). It was later found that bees can see ultraviolet light, and that some flowers have very distinctive ultraviolet patterns. Insects can, in addition, be sensitive to the plane of polarized light.

Color also plays a part in stimulating sexual behavior—for example in sticklebacks *(Gasterosteus aculeatus)*. The male changes color at the mating season—his belly becomes red and his back turns a blue-white. This coloration attracts females, who, by now, are plump with eggs. The male guides a female to a nest that he has constructed. She enters it and he nudges her from behind, causing her to release the eggs. She then swims away, and the male enters the nest and fertilizes the eggs.

Many visual stimuli to which animals respond are crude. For example, a robin will attack a clump of red feathers on its territory during the breeding season, and some cuckoos automatically feed a gaping mouth whether it is that of a fish or a fledgeling. But some stimuli are supernormal; these are releasers which evoke a greater response than the natural releaser — such as egg size and color. Oyster catchers (family Haematopodidae) and other birds, for example, will try to brood an egg considerably larger than their own in preference to their natural egg.

Animals often respond preferentially to different stimuli. Herring-gull chicks *(Larus argentatus)*, for instance, prefer a high level of contrast on a bill at which they peck to elicit food from the parent. When hungry they peck at a red patch on the lower mandible of the parent's yellow bill, which causes it to regurgitate food. But it has been found that a red and white striped pencil is a better releaser than a real gull bill. This is a case of misfiring, when the animal's behavior is elicited under inappropriate circumstances and does not attain its goal.

Scent organs are essential for most animals in the recognition of scent trails to food sources, but scent is also important in communicating fertility. At the time of estrus, many female mammals secrete odiferous pheromones which are detected by the males.

Parental behavior

Different animals rear young in very different ways. In some ducks, the male deserts the female to mate with others in that season. In sticklebacks, however, the female leaves the male to brood the eggs and care for the young fish until they become independent. But among birds and mammals both parents often care for the young which may be unable to feed themselves or maintain an adequate body temperature for several weeks.

Ties between parents and their offspring are not necessarily exclusive. Among prairie dogs *(Cynomys ludovicianus)*, for example, which live in communities for most of the year, the young are suckled indiscriminately by any lactating female, and are groomed by any male.

Feeding

The life styles of all animals are organized around their methods of obtaining food. Some parasites, for example, modify the behavior of an intermediate host for the next stage of their life cycle. Many parasitized fish swim close to the surface of the water, unlike those free from parasites, making it more likely that they will be caught by fish-eating birds and thus pass the parasites on to another host.

Many animals feed in groups, which has the advantage of security. It is more likely that a predator will be detected by one member of the group sooner than if the animals are feeding separately. In addition, animals grazing in a group look up less than when they are on their own, which means that each individual can spend more time eating.

Social behavior

Some animals are solitary all the time, and others are solitary most of the time but come together for certain activities, for example to migrate. A species of locust *(Locusta migratoria)* has a solitary phase, when its coloring is green. But when environmental conditions are favorable, the insects change color to black and russet and become part of a migratory swarm. Some birds, such as the Eurasian robin *(Erithacus rubecula),* will attack any member of their own species that comes near their territory, but will join a flock which migrates as winter approaches.

Other animals collect together and coordinate their activities. Termites, ants and bees, for example, live in tightly regimented societies or colonies. The division of labor between the different animals in the colony is strict, and the number of different kinds of individuals depends on what the colony requires for its survival. The colonies are controlled by chemicals called pheromones which are produced by the queen. These chemicals and the high degree of physical contact between members of a hive ensure cohesion.

Most primates live in groups. These are less regimented than those of social insects but order is maintained by a social hierarchy, ruled over by a single dominant male, usually the oldest and largest. This male has the first choice of the best food and most receptive females. Aggressive confrontations are usually avoided by the use of a large number of facial expressions and dominant or submissive gestures.

Animals that live in close communities often hunt together. Lions, for example, hunt by stalking and then ambushing their prey. But hunting

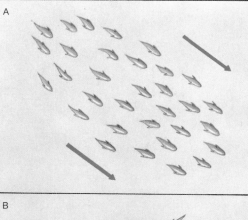
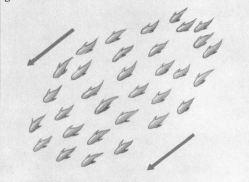

Schooling fish are remarkable in the way they travel and communicate in shoals. The shoals are elliptical in shape (A) and contain fish of roughly the same size, which travel at equal distances away from one another, at the same speed. There does not seem to be a leader and a change of direction is made by all the fish at the same time (B). This indicates a communication between the fish, but it is not fully understood by zoologists.

dogs run down their prey in packs, the average size of which is six males and two females. There is usually no dominant dog, male or female, although a dog will determine the direction of movement of the group. The dogs return to the lair after the kill, where the pups wait, and the whole pack is fed on regurgitated meat.

Animals such as hunting dogs, which have a high-protein diet, can afford to spend a lot of time in activities such as play because the energy value of the food they consume is high. Play involves romping and mock fighting and is important to young animals because through play they learn skills that they will need as adults.

Grooming is practiced by most mammals but among primates, particularly Old World monkeys, it reflects rank order and serves to maintain the hierarchy of the group. Subordinate individuals spend more time grooming dominant clan members than being groomed themselves. Male baboons *(Chaeropithecus* sp.), for example, lie in front of females, which are expected to groom them. The males groom back, but for not as long. Grooming is also an indication of reassurance and friendliness.

Invertebrates

Of the million or so existing animal species, more than 95 per cent have no backbone and are classed as invertebrates. They have a wide variety of shapes and sizes, from microscopic single-celled (acellular) animals only a fraction of a millimeters in length, to the giant squid *(Architeuthis* sp.), which may measure up to 55 feet (17 meters) long and is strong enough to fight with whales that prey upon it. Invertebrates have an extraordinary range of different life styles and occur at every level of — and play a vital part in — the complex food web that links all forms of life.

The range of invertebrates
There are about 33 invertebrate phyla, but because new groups are continually being discovered and the classification of invertebrates is constantly under revision, the exact number and distribution of phyla is debatable.

Nine groups are particularly important because they represent about 90 per cent of all living invertebrates. From the simplest form to the most complex these nine phyla are: acellular animals (Protozoa); sponges (Porifera); jellyfish and their allies (Cnidaria); flatworms (Platyhelminthes); roundworms (Nematoda); snails, squids and clams (Mollusca); earthworms, rag-

worms and leeches (Annelida); insects, spiders and decapods (Arthropoda); and starfish and sea urchins (Echinodermata).

The protozoa are all very small, ranging from three millimeters in length to a microscopic size consisting of a single cell, within which all the functions of life are carried out. For this reason many scientists classify protozoa in their own subkingdom to distinguish them from the multicellular animals, which are classed as Metazoa. The simplest multicellular group is the sponges, with some 10,000 species.

The cnidarians include the jellyfish, sea anemones, corals and hydroids. They are one of the most simply organized of the multicellular animal groups, having two layers of cells which surround a tubular body cavity, with an opening at one end forming a mouth, and no specialized tissues for respiration or excretion. Most cnidarians are marine animals, although there are a few freshwater species. The platyhelminths fall into three classes — the free-living turbellerians, the parasitic flukes (Trematoda) and the parasitic tapeworms (Cestoda). They have developed a little further than the cnidarians, having three layers of body cells, but most of them lack a body cavity, a circulatory and an excretory system, these func-

The zebra spider *(Salticus scenicus)* is a jumping spider and a member of the phylum Arthropoda — the largest group of animals, containing about 900,000 species. Jumping spiders can leap a short distance; they hunt primarily by sight and have the best vision of all spiders. Their eight eyes are arranged in two rows of four; they consist of a large number of receptors and can perceive a sharp image of considerable size. In addition to their keen eyesight, the spiders have chemosensitive hairs at the tips of their appendages.

tions being carried out by cells on the surface of the body. The turbellarians and flukes have a digestive tract which opens from the mouth. The Cestoda, however, absorb food directly from their hosts and have evolved with no digestive tract at all.

The roundworms, or Nematoda, are unsegmented and usually have a digestive tube with a mouth at one end and an anus at the other. The gut is surrounded by a fluid-filled cavity — the pseudocoel — which acts as a hydrostatic skeleton. They are an extremely successful group and are found free-living in all environments, although most of the 10,000 known species are parasitic. Nearly all vertebrate animals and many invertebrates can be hosts to parasitic roundworms.

The annelids, or segmented worms, are the most highly developed worms. They are divided into three major groups: bristle worms (Polychaeta), most of which are marine; earthworms (Oligochaeta), which are mainly freshwater and terrestrial; and leeches (Hirudinea), which are found in the sea, freshwater and on land.

The phylum Echinodermata includes starfish, sea urchins, sea lilies, sea cucumbers and brittle stars. These five distinct groups are based on the same radial structure, usually consisting of five or ten arms radiating from a single mouth. They are exclusively marine and are all slow-moving or sessile.

The phylum Mollusca is the largest after the arthropods, containing more than 100,000 species, and probably the most sophisticated of all

invertebrates. Most mollusks have a hard shell enclosing a soft body and although they conform to a general anatomical pattern, their external body forms are extremely varied. The mollusks contain three main classes — Gastropoda (such as snails); Bivalvia (which contains oysters, mussels and clams); and Cephalopoda, which includes octopuses, cuttlefish and squids.

The arthropods are the most successful invertebrates in terms of numbers of species (about 900,000). All have jointed limbs and segmented bodies, covered by a hard exoskeleton. This body covering acts as a protective armor and as a frame onto which muscles can attach. The largest class of arthropods are the insects (Insecta), which are primarily land-dwelling. Crustaceans, which are found mainly in salt and freshwater, are the second largest class of arthropods, and show a greater diversity of structure and physiology than the insects. They range from the relatively advanced lobsters, crayfish and crabs, to the tiny copepods that graze on the phytoplankton in the oceans. The third group comprises the arachnids — the spiders, ticks, scorpions and their allies — which are mostly terrestrial and usually have four pairs of legs, no wings or antennae, and pincers rather than simple mouths. Of the six remaining classes of invertebrates, only the centipedes and the millipedes are represented in great numbers.

The minor phyla
The remaining phyla include the smallest phylum Placozoa — with only a single species — and Priapulida, which contains only nine known species of tiny, cucumber-shaped, seabed-dwelling worms. Most of the small phyla comprise marine species which are sand or seabed dwellers.

Some invertebrates seem to form a transition between the invertebrates and the vertebrates. These are: the hemichordates (such as acorn worms), the tunicates (including sea squirts) and the cephalochordates (lancelets), which are members of the phylum Chordata and have a simple skeletal rod (or notochord) at some stage of their development.

Sea slugs (order Nudibranchia) are gastropod members of the phylum Mollusca, which also includes snails, limpets, abalones and slugs. Among the most brilliantly colored of all marine invertebrates, they have no shell (unlike most other mollusks). Some have colored tentacles called cerata along their dorsal surface.

Protozoa and sponges

Stentor sp. is a large ciliate, about one inch (2.5 centimeters). The cilia around the rim of its trumpet-shaped body are joined to form membranelles, or plates, which beat to create feeding currents. It is a selective suspension feeder, rejecting those food particles that are of no nutritive value to it.

The Protozoa are divided into four subphyla: the Sarcomastigophora (amebae and flagellates), the Ciliophora (ciliates), the Sporozoa and Cnidospora. They are acellular organisms and constitute the simplest animal group. Most of them are microscopic. They comprise about 20,000 species and are found in almost every habitat where moisture is present — in marine, freshwater and terrestrial environments — and also as parasites in most animals. Because of their resistant spores, some protozoa can withstand extremes of temperature and humidity. It is not surprising, then, that they are among the most numerous animals in the world. Some protozoa share similar features with certain sponges, which suggests that these two groups may be related.

Amebae and flagellates

Amebae have a constantly changing body shape and move by producing pseudopodia (false feet). The cytoplasm of the ameboid cell is extruded at one point to form the pseudopodium as the animal moves forward. The common *Amoeba* has a naked cell surface, but a variety of shelled forms exists; the genus *Difflugia*, for example, constructs a case from minute grains of sand, whereas other amebae secrete intricate shells of calcium carbonate and silica. These shells contain holes through which the pseudopodium can be extruded to collect food particles.

Some amebae are parasitic, such as *Entamoeba coli* which lives in the human large intestine. Found in up to 30 per cent of the world population, it does not transmit disease, but scavenges bacteria and food detritus. Such a nonharmful association is termed commensalism. But the related *Entamoeba histolytica* is harmful and causes amebic dysentry.

In contrast to amebae, flagellates move using a long hairlike structure called a flagellum, which beats like a whip to provide propulsion. Most flagellates have a fixed body shape (usually oval) and almost all reproduce asexually by binary fission, but there is considerable diversity between species. Choanoflagellates, for example, have a delicate collarlike structure which surrounds the base of the flagellum and helps to collect food particles. Similar collar cells (choanocytes) are found in sponges, which suggests a possible link between these two groups. In some collared flagellates the individuals do not separate after cell division, but give rise to a colony.

Most flagellates feed on small particles and organic materials dissolved in the surrounding water. But some, such as the green *Euglena* commonly found in ponds, are able to produce their

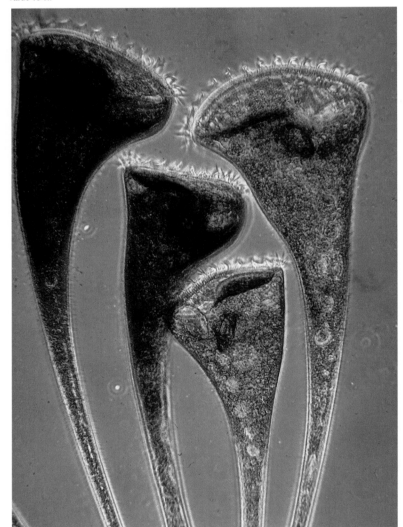

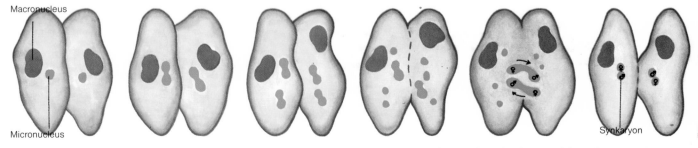

Macronucleus

Micronucleus

Synkaryon

The ciliate *Paramecium* reproduces by conjugation. When two adults with compatible micronuclei come together, the micro- nucleus of each divides twice. Three of the four new cells disintegrate but the fourth continues to divide. The cell walls join- ing the two ciliates dissolve and the male nuclei of the two conjugants are exchanged. The adults then separate, and development of the synkaryon continues in each one. At about this stage the macronucleus starts to disintegrate and is resorbed. The synkaryon own food by photosynthesis. The presence of the photosynthetic pigment, chlorophyll, has led many biologists to classify such flagellates as algae (simple plants). In all other respects, however, they are identical to flagellate protozoa.

Many flagellates live in association with other animals. *Trichonympha*, for example, lives in the intestine of termites. The termite relies on the protozoon to digest the wood that it eats — a mutually beneficial relationship known as symbiosis. But some flagellates are blood parasites, such as *Trypanosoma*, and cause disease (in this case sleeping sickness).

Ciliates

The most complex and diverse species of protozoa owe their name to the cilia (short hairlike fibers similar in structure to flagella) which grow in orderly rows on the body and beat rhythmically to propel the animal. In many ciliates the cilia occur only on parts of the body, whereas in others they form plates or "membranelles." Still others have cilia which are fused into stiff cirri and used as legs for crawling. Ciliates reproduce asexually by binary fission, or sexually by conjugation. They differ from other protozoa in that they have two nuclei — a macro- and micronucleus (other protozoa have only one).

Sporoza and Cnidospora

These organisms, the spore-formers, have no distinct locomotory adaptations because they are all parasitic. They live in all animals and are often transmitted by insect vectors. Their name comes from the production of spores, or cysts, during the infective stages of their life. The life cycle of this group is complicated — reproduction alternates from asexual to sexual. Asexual reproduction involves the binary fission of spores, usually in the host. The offspring develop into gametes which mature and fertilize other gametes which eventually produce spores.

Sponges

Sponges comprise the phylum Porifera (pore-bearers) and represent the simplest level of multicellular animals. Most of the 10,000 species are found in shallow waters, although some live in deep water. They range in length from one-quarter of an inch to more than four feet (0.5 centimeter to 1.2 meters).

In contrast to the colonial protozoa, sponges consist of several cell types, each of which performs a specific function and is independent. This feature means that regeneration is easy — if a sponge is fragmented, the cells simply organize themselves into a new sponge.

The simplest sponges are tubular with an external layer of epithelial, or lining, cells. The internal surface is covered with flagellated collar cells (choanocytes) which maintain a water current through the sponge. Food particles are extracted from this current which is also used for gas exchange and waste removal. Water is drawn in through small pore cells in the walls of the sponge, and ejected from the large mouth (osculum). More advanced sponges have complex systems of canals and chambers, through which water is channeled.

Sponges reproduce sexually and asexually. Most are hermaphroditic, and produce eggs and sperm at different times, but some are dioecious. Asexual reproduction occurs by budding. In addition some sponges produce gemmules which survive when the parent body disintegrates in winter. In spring the gemmule develops into an adult sponge. Sexual reproduction occurs when sperm is released and enters another sponge in the water currents.

Glass sponges, or hexactinellids, comprise the class Hexactinellida, and are mainly deepwater sponges. Like other sponges they have special cells which secrete the skeleton which is made up of spicules, or tiny needle-like structures. In bath sponges, these spicules are composed of a horny protein called spongin, but in glass sponges they are made of silica. These siliceous spicules are fused to form a six-pointed shape from which the class name of the glass sponges is derived.

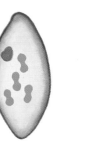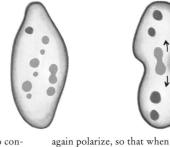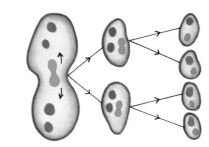

divides three times producing eight new cells. Three of these dissolve, four start to attain macronucleus status and one continues to

divide as a micronucleus. The ciliate itself then starts to divide. The four macronuclei polarize, two at each end, and when the cell

finally divides it also contains one micronucleus. The micronucleus in both new ciliates continues to split and the macronuclei

again polarize, so that when the two ciliates divide, the resultant four cells each contain one micro- and one macronucleus.

Coelenterates

The coelenterates are among the simplest of the metazoans (multicellular organisms with cells organized into tissues), and include those animals in the phylum Cnidaria, such as corals, jellyfish, hydras and sea anemones, and the comb jellies of the phylum Ctenophora. The largest species is the jellyfish *Cyanea artica* — 7½ feet (2.3 meters) across with tentacles 100 feet (30 meters) long — and the smallest are the individual polyps of some coral colonies, which measure from one millimeter in length. Coelenterates comprise about 9,000 species and are probably the most common macroscopic marine animals, especially in tropical and subtropical coastal waters. Furthermore, the corals create reefs and atolls, which are some of the world's most diverse and productive habitats.

The sea nettle *(Chrysaora hyoscella)*, common in the coastal waters of the Atlantic Ocean and Mediterranean Sea, has a diameter of 12 inches (30 centimeters). Its fringing tentacles are covered with nematocysts, as are its four trailing arms, which presents a danger, although not a fatal one, to swimmers.

In the reproductive cycle of most jellyfish the egg released by the adult medusa becomes a ciliated planula larva which develops into an attached polyp. When medusoid layers bud off the polyp it is called a scyphistoma and when the young medusae (ephyrae) are mature enough to detach themselves, it is known as a strobila.

General structure

Cnidarians have two basic forms — polyps and medusae — which are radically symmetrical, with no definite front or rear ends. The flowerlike polyps are attached at their base to their mother organism, and have a mouth on the upper side, surrounded by tentacles. Medusae, commonly known as jellyfish, are free-swimming, bell-shaped organisms, with a mouth on the underside. In the typical life cycle of a cnidarian, the two basic forms alternate — polyps bud off medusae, which then produce a larval polyp, and so on. One form is usually dominant, however, and in some species the other is omitted altogether.

The coelenterate body comprises two layers of cells — an outer epidermis and an inner gastrodermis — separated by a layer of jelly-like matter: the mesogloea. The two layers of cells surround a central cavity (coelenteron) and contain muscle cells which contract to bend the body or retract the tentacles. In cnidarians the epidermis contains stinging cells (nematocysts) to immobilize prey. The food is then pushed by the tentacles through the mouth into the coelenteron, where it is digested by enzymes and absorbed by the gastrodermis.

Many coelenterates often form colonies of thousands of individuals. In the simplest colonies all individuals are identical, but feed and reproduce separately. In more advanced colonies, however, such as those of some corals and hydrozoans, several different forms are present (polymorphism), and each type undertakes a different function. But in many there are nervous connections between individuals — if one polyp is touched, the whole colony contracts.

Hydrozoans

Some members of the class Hydrozoa, such as the common freshwater *Hydra*, have no medusa stage although many other members of the class develop both polyps and medusae, and some display the medusoid form only. In the reproductive cycle of most hydrozoans, sexually reproductive medusoids bud off the parent asexually and either drift free or remain attached to the parent organism (when they are known as gonophores). In the *Hydra*, however, reproduction is more commonly asexual, by budding from the parent.

Most hydrozoans are colonial and, although normally attached, move by a creeping action at the base. Most of them are green, because of the

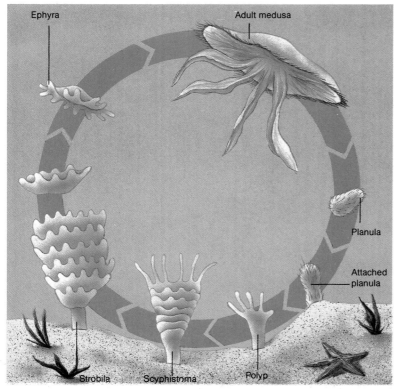

Ephyra

Adult medusa

Planula

Attached planula

Strobila

Scyphistoma

Polyp

presence of symbiotic algae (zoochlorellae) in the gastrodermal cells.

The best-known hydrozoan, the Portuguese man-of-war (*Physalia* sp.) has a common hydrozoan feature — an air-filled cushion which acts as a float. Its tentacles comprise a colony of polymorphic polyps.

Jellyfish

Jellyfish (class Scyphozoa) consist of a swimming bell fringed by tentacles, with a four-cornered mouth on the underside. Around the mouth are four trailing arms which, like the tentacles, are well supplied with nematocysts. Jellyfish swim by contracting and releasing a ring of muscle cells; balancing organs (statocysts) and simple light receptors around the bell margin help them to remain upright. As their name suggests, they contain large amounts of the jelly-like mesogloea, which helps to control buoyancy, and also acts as an elastic support for the body.

Sea anemones, corals and ctenophores

The sea anemones and corals of class Anthozoa (flower animals) have a polyp stage only. These are often large and complex, with the coelenteron divided by partitions, or septa. Many sea anemones have large attachment disks and thick, leathery bodies, which allow them to survive on rocks which are exposed at low tide.

Corals are essentially polymorphic colonial sea anemones, although solitary forms do exist. The polyps secrete skeletons, the exact form of which defines the species, such as the delicate sea fan (*Gorgonia* sp.). Reef-building corals flourish only in tropical and subtropical coastal waters where temperatures are above 65°F (18°C) and the water is less than 100 feet (30 meters) deep. The necessity for shallow water seems to reflect the light requirements of the photosynthetic algae (zooxanthellae) which live in the gastrodermal cells. The coral skeleton is built up faster in species with zooxanthellae than in those without them, possibly because algal photosynthesis increases the production of calcium carbonate (which creates the skeleton) by removing carbon dioxide.

Ctenophores, or comb jellies, are grouped with cnidarians because they are also radially symmetrical, jelly-like, and composed of two layers of cells. But the gastrovascular cavity of the cnidarians has become a canal system in the ctenophores and the medusoid shape only has been adopted and modified into a sphere or oval. There are two classes: those with tentacles, such as the most common form, the sea gooseberry *(Pleurobrachia pileus)*, and those without. These animals are marine and are usually found swimming among plankton; many are luminescent. They move using ciliated bands called comb plates. The cilia fuse into plates during the development of the animal; by beating these plates consecutively from the head to the tail, they swim through the water. Ctenophores catch their prey using lasso cells (colloblasts) which have a similar function to that of nematocysts in cnidarians, but they are adhesive rather than paralyzing.

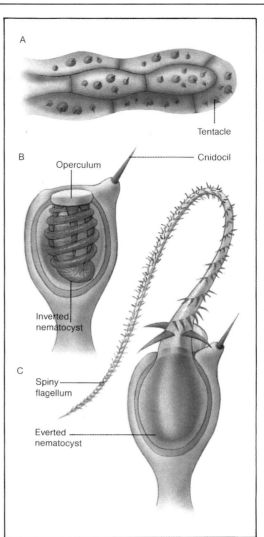

A

Tentacle

B

Operculum

Cnidocil

Inverted nematocyst

C

Spiny flagellum

Everted nematocyst

Nematocysts are abundant on the tentacles of coelenterates (A). These oval cells contain an inverted tube at one end of which is a flagellum, often covered with spines, coiled tightly round the bulb of the tube (B). On the activation of a protruding hair-like trigger (cnidocil), the cell lid (operculum) snaps open, the nematocyst whips out everting itself (C), and the toxins contained in it are ejected through the spines.

Sea anemones, like most coelenterates, are carnivorous, the larger ones feeding on small crustaceans. The nematocysts in their tentacles do not have a cnidocil, nor an operculum; they often simply break through the capsule wall. The tentacles surround an oral disk which contains a mouth. Prey is carried to the mouth, which is widened by muscles, and the prey is swallowed whole.

Platyhelminthes

The phylum Platyhelminthes (flatworms) is divided into three classes: the turbellarians, the flukes (Trematoda) and the tapeworms (Cestoda). Only the turbellarians are free-living; the other two are exclusively parasitic. As their common name suggests, flatworms are flattened, soft-bodied organisms. Most flukes are microscopic, but intestinal tapeworms may grow up to 30feet (9meters).

General features
Compared with the simpler coelenterates, flatworms show several advanced characteristics: for example, the body is bilaterally symmetrical, and has a definite head end. Free-living flatworms are active animals and in many species the head carries pairs of eyes, as well as chemoreceptors. There is also a concentration of nerve cells at the front end of the body which forms a primitive brain, in addition to the diffuse nerve net similar to that of the cnidarians.

The flatworm body is composed of three layers of cells, whereas in coelenterates there are only two. Between the epidermis and the gastrodermis (which lines the digestive cavity) there is a layer of mesodermal cells. Oxygen reaches this middle cellular layer by diffusion because these animals are so highly flattened that a specialized breathing system is not required. The mesoderm contains the complicated reproductive organs which are made up of different types of cells. This differentiation represents a higher level of organization than that found in the coelenterates, which possess tissues, but not organs.

As in *Hydra*, the digestive cavity in flatworms (apart from tapeworms) has one entrance only and no separate exit, with the result that undigested food is ejected through the mouth. In turbellarians this digestive cavity is often highly branched, so that food can be distributed to all parts of the body.

Turbellarians
The turbellarians include all free-living flatworms. Some species grow up to 25inches (60centimeters) but most are less than 2inches (5centimeters) long. Most freshwater species are drab and inconspicuous, whereas tropical marine species may be very colorful.

Turbellarians do not swim freely, although they live in water, but creep along the bottom. The epidermis is equipped with minute, hair-like cilia and glandular cells which secrete mucus in which the cilia beat, enabling the flatworm to glide along. They also have bands of muscles which allow complex bending movements of the body.

Most turbellarians are carnivorous, feeding on small animals, or necrophagous (feeding on dead animals). Many secrete mucus and an adhesive to entangle their prey, before using the protrusible pharynx to break it up into small particles. In some cases the prey is ingested whole.

Compared with parasitic flatworms, turbellarians have a simple life cycle. Most species are hermaphroditic, with male and female sex organs in the same individual. Self-fertilization does not usually occur, however, and a partner is needed for fertilization to take place. Sperm is exchanged between the two individuals and both lay fertilized eggs in egg capsules or gelatinous masses.

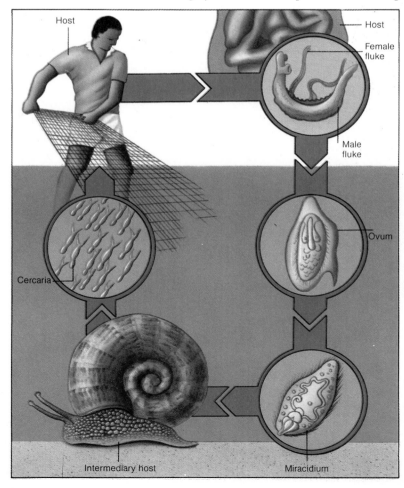

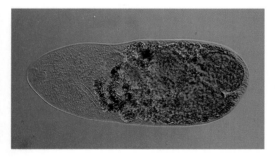

The blood fluke *Schistosoma mansoni* (left) breeds in the host's intestine or bladder. Fertilized eggs pass out in the feces, develop into larvae (miracidia) and are released in water. In an intermediary host they breed cercariae which burrow into a third host.

The small intestinal fluke *Metagonimus yokogawei* (above) is one of many that are found particularly in Asia. This species has only one intermediary host, usually snails or fishes. It is transmitted to vertebrate hosts which eat raw fish and mollusks.

After two or three weeks, young flatworms resembling their parents hatch from the eggs.

Some turbellarians also reproduce asexually; the body constricts in the middle and the hind portion breaks off to form a complete new individual. Using the same principle, if an individual is cut in two, the front half will form a new tail and the rear half a new head.

Flukes

These flatworms occur as internal parasites (order Digenea) and external parasites (order Monogenea). They generally have leaf-shaped bodies with suckers or, rarely, hooks to attach to their hosts. As a result of their parasitic habits, adult flukes lack the sense organs and layer of cilia that are found in turbellarians.

Another feature found in flukes and typical of most parasites is a high reproductive ability, which ensures that at least some offspring reach suitable hosts. Not only do the adult flukes of internal parasites produce large numbers of eggs, but the larvae of these flukes also reproduce within their hosts.

External parasites are often found in the gill cavities of fish, where they have to withstand strong water currents. They feed on blood and gill tissue. Internally parasitic flukes may be found in many vertebrates as well as in other invertebrates. The liver fluke *(Fasciola hepatica)* is a well-known parasite found in the bile passage of the liver of sheep and cattle where it inflicts fatal damage. This fluke has a complex life cycle which involves several hosts. The eggs leave the vertebrate host in the feces and hatch in water into free-swimming larvae (miracidia). These larvae swim until they encounter a snail — the intermediate host — into which they burrow. Inside this host the larvae multiply and eventually leave the snail as another larval form — cercaria. Up to 600 cercariae may be produced from a single miracidium. These then encyst on grass where they are eaten by the vertebrate host, in which they develop into sexually mature flukes.

Tapeworms

The adults of this parasitic group are found as internal parasites in the alimentary canal of vertebrates. Like the flukes, tapeworms have no obvious sense organs and their complex life cycle usually involves two more hosts. The beef tapeworm *(Taenia saginata)*, for example, has two hosts — cattle and man.

Tapeworms have no digestive system — indeed, they do not need one, because they are surrounded by digested food in the gut of their host and they simply absorb nutrients through their cuticle. These worms consist of a head, or scolex, from which segments or proglottids bud off. The head usually bears hooks and suckers for attachment to the gut lining of the host. The proglottids remain attached in a long chain as they mature, finally breaking off and leaving the host in the feces. The eggs hatch from the proglottids once they have been eaten by the intermediate host.

The candy-striped flatworm *(Prosthecereaus vitatus)* is a member of the order Polycladida. The order name refers to the many-branched intestinal system with which the organism digests its food. It is marine and reaches 2 inches (5 centimeters) in length. Gland cells below the epidermis secrete an adhesive substance which allows the flatworm to stick to a substrate or to prey. The combination of this substance with beating cilia and muscular contractions enables the flatworm to move over any surface.

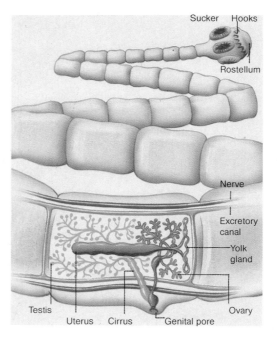

The reproductive segment (proglottid) of a tapeworm contains male and female reproductive organs. During copulation, the cirrus of the male is inserted into the genital pore of another proglottid and the fertilized eggs are stored in the uterus. Cross-fertilization usually occurs with a proglottid from another worm in the same host or with another proglottid on the same worm if the worm has twisted around itself. The proglottid breaks off the parent organism and when the eggs are mature the proglottid wall ruptures, releasing them.

Nematodes and annelids

In the evolution of metazoans the stage of development after that of the three-layered structure of the flatworms seems to be the appearance of a fluid-filled cavity, or coelom. This body cavity occurs in two forms: the pseudocoelates — such as the aschelminths, which include the gastrotrichs, rotifers, kinorrhynchs, nematodes and nematomorphs — and the coelomates, such as the annelids. Another feature which appears at this stage is metamerism, or segmentation, displayed particularly in the annelids. All of these animals share some features but are too different to be placed under one phylum, so each has its own.

Ribbon worms

This group, known as nemerteans, are not pseudocoelate because a solid tissue (parenchyma) fills the cavity between the gut and the body wall.

In this respect they resemble the flatworms but are more highly specialized because they have more complex nervous and circulatory systems, and an alimentary canal with a mouth and anus.

Most of the 600 or so species are marine burrowers and feed on invertebrates. They have a remarkable food-catching apparatus — an eversible tube (the proboscis) which is shot out by hydrostatic pressure through a pore in the head. In many species the proboscis simply coils around the prey but in some it has teeth and stabs prey.

Roundworms

This group, the nematodes, is the largest of the aschelminths, containing about 10,000 species. They are found in most environments, and millions may occur in only a couple of acres of soil.

Nematodes have cylindrical bodies, which taper at each end and are covered in a thick protein layer called the cuticle, which is shed periodically as the animal grows. The cuticle and body wall are permeable to water which flows in and out continuously. The intestine of these organisms runs from the mouth to the anus, and is enclosed by longitudinal muscles along the length of the body. Roundworms move like snakes by contracting these muscles, helped by the elasticity of the cuticle and the pressure of the fluid in the pseudocoel. They have a simple nervous system, as do most aschelminths; the brain is situated anteriorly and comprises a ring of ganglia from which nerves run down the length of the body.

Nematodes have separate sexes and give birth to larvae which resemble the adult. Many are parasitic — some only in the larval stage, others only when they are adult, and still others are parasitic throughout their life. Parasitic nematodes, such as *Dorylaimida* sp., damage crops by sucking the contents from punctured plant cells or by feeding on the tissues inside the plant.

Nematodes also parasitize animals, including humans. The hookworm (*Necator americanus*) does great harm by feeding on the blood and cells of the intestinal lining; female hookworms produce many eggs which leave the host's body in the feces. In unsanitary living conditions the eggs hatch and the young hookworms enter the human body by boring through the skin of the feet.

Annelids

Annelids are worms in which the body is divided into many segments, and as such are a little more developed than the pseudocoelates, which are unsegmented. The phylum Annelida has three classes: Polychaeta, Oligochaeta and Hirudinae. In the polychaete ragworm (*Nephthys caeca*) each segment of the body — apart from the head and the last segment — is identical, and the external and internal organs are repeated in each segment. In the oligochaetes, such as the earthworms, and the leeches (class Hirudinea) some segments are specialized for particular functions

The nematode *Ascaris* sp. parasitizes the intestine of humans, cattle and pigs as well as other invertebrates. The adult worms breed in the intestine and the eggs develop in cysts in the feces. When this excrement is taken in by another host, the eggs hatch in and break through its small intestine. Once in the bloodstream they are carried to the lungs, make their way to the mouth and are swallowed. The young worms then burrow into skeletal muscle, remaining there until the flesh containing them is eaten by another host.

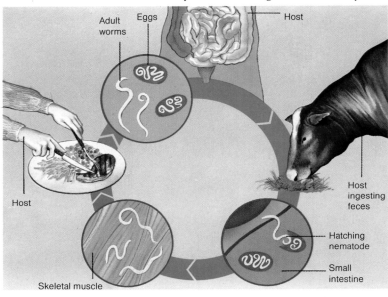

Adult worms — Eggs — Host

Host

Host ingesting feces

Hatching nematode

Small intestine

Skeletal muscle

Rotifers are so called because the head region of these organisms has a crown of cilia which beat very fast to propel it, and in some species looks like a spinning wheel. This species, *Philodina gregaria*, retracts its corona as it creeps along the substrate, and uses the two spurs at its foot as a means of attachment. The white oval shape in the head region is the mastax, or pharynx, which is used to catch and break up food. The orange area is the stomach. This species reproduces by parthenogenesis only and has no males.

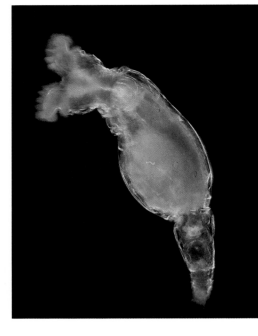

and are not all identical. Like nematodes, annelids have a space in the body surrounding the gut but, because it is formed in a different way during the development of the embryo, it is called a coelom.

Polychaete annelids are marine. Some, such as the ragworm, are fast-moving predators, with a well-developed head region that has eyes, sensory tentacles and powerful jaws. They move by using the fleshy leglike extensions of the body (parapodia) on each segment. Other polychaetes such as lugworms *(Arenicola* sp.) burrow in mud; still others are sessile and live in tubes, as do the fanworms *(Sabella pavonina).*

Most of the 3,000 or more species of oligochaetes are burrowers and live in freshwater and terrestrial habitats. Instead of parapodia, each body segment has a few stout bristles (setae). These worms travel by peristaltic contractions, as do burrowing polychaetes, using the bristles as a temporary anchor.

As in polychaetes, the reproduction of oligochaetes is usually sexual, but they contrast with the polychaetes in that they are hermaphroditic, each worm having both male and female sex organs. These are situated toward the front end of the animal in different segments. During copulation earthworms come together in opposite directions with their undersides pressed together. They are held together by a mucous tube which is secreted by the glands in the clitellum (an enlarged segment girdling the body). The clitellum of each worm attaches itself to the segments containing the spermathecae of the other worm. Sperm is moved to the clitellum by muscular contractions and passes through a groove in the clitellum into the spermathecal opening of the opposite worm. After a few days the clitellum secretes a hard ring of material which slips forward and collects eggs and sperm as it moves over the genital openings. When it comes off the worm, the ends seal up; this cocoon can carry 20 eggs which hatch after about 12 weeks.

Leeches are found in fresh water, the sea and on land. They share several features with oligochaetes, including hermaphroditism, fertilization in a cocoon secreted by a clitellum and the lack of parapodia, but the coelom has been lost during secondary adaptation, except in one species.

Leeches have flattened, muscular bodies with a sucker at each end which is used as an anchor while they move by extending and contracting the body. Most leeches are ectoparasitic, feeding on the surface of their host. The front sucker is clamped onto the victim's skin and the teeth in the leech's mouth then make a small cut, which is unnoticed by the host due to an anesthetic which is secreted by the leech. While the leech pumps blood out of the wound, it secretes a chemical to prevent the blood from clotting. Bloodsucking leeches feed infrequently, but when they do feed they can draw out several times their own weight in blood in one meal.

The tubeworm *Spirographis spallanzani* is an annelid which inhabits the Mediterranean Sea. It lives in a tube which is made of sand cemented by an organic material that it secretes from special glands, and is constructed on the sea bed. The movement of small cilia on the tentacles of the worm creates water currents which carry food toward its mouth.

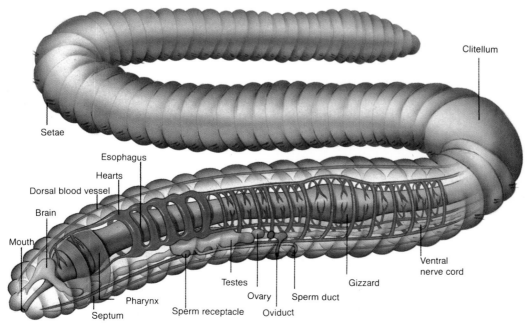

Setae
Esophagus
Hearts
Dorsal blood vessel
Brain
Mouth
Pharynx
Septum
Sperm receptacle
Testes
Ovary
Oviduct
Sperm duct
Gizzard
Ventral nerve cord
Clitellum

Earthworms diffuse gases and water through the cuticle although oxygen is also distributed by hemoglobin in the blood. These animals have five hearts from which branched vessels carry blood along the length of the body. The nervous system of earthworms is well developed compared with that of the flatworms: two cerebral ganglia form a brain from which two nerve cords run and although the worm has no eyes, it has photoreceptors in the epidermis particularly at its front and back ends.

Echinoderms

The phylum Echinodermata (spiny-skins) consists of about 5,000 species of exclusively marine organisms which include such well-known animals as starfish and sea urchins. The description "spiny-skins" is derived from the calcareous (chalky) plates that form the endoskeleton of these animals which in some species is covered with spines. These spines are particularly prominent in sea urchins.

The most obvious feature of these animals, however, is their five-rayed, or pentamerous, symmetry. It is surprising, then, that the rich fossil record of echinoderms (made possible by the presence of the skeletal plates) shows that some of the earliest species were actually bilaterally symmetrical. This evidence, plus the fact that modern forms pass through a bilaterally symmetrical larval stage, suggests that radial symmetry is a secondary adaptation.

The relationship between the echinoderms and other invertebrate phyla is somewhat obscure. It was initially thought that echinoderms and hemichordates (primitive chordates) were closely related because of the similarity between the larvae of the two groups. It is now believed, however, that the resemblance is due not to homologous (genetically similar) characteristics but to the adaptation of different organs to perform the same function (a process that is known as convergent evolution).

Starfish

Starfish, or sea stars (class Asteroidea), are the most familiar echinoderms. They consist of a central disk from which five arms arise (although some species have more than five arms). These animals move using tube feet (podia) which are found in a groove (the ambulacral groove) under each arm. Each tube foot ends in a sucker which allows the starfish to stick to rocky surfaces; when moving over sand, however, the tube feet are used as stiff legs.

Tube feet are also used to deal with prey which usually consists of bivalves such as clams and oysters. The tube feet exert enough pressure to open them slightly so that the starfish can push its now everted stomach into the gap (the stomach can squeeze through a space as small as 0.1 millimeters). The stomach then secretes enzymes which partly digest the victim before it is eaten. The starfish intestine has five branches, one in each arm. Indigestible fragments are usually ejected via the mouth (there is an anus on the upper surface but it is seldom used).

Within each arm there is also a pair of gonads (sex organs). Reproduction in most echinoderms is a simple process. The sexes are separate and sperm and eggs are released into the water, where fertilization occurs. The larvae that hatch are called bipinnariae and metamorphose gradually into the adult form. A few species copulate and the eggs develop in a brood chamber without going through a larval stage. Starfish are also able to reproduce by fragmentation — or regeneration, as this process is sometimes called — but this is only an occasional occurrence.

In large numbers, starfish can have a devastating ecological effect. The population of the crown-of-thorns starfish *(Acanthaster planci)*, for example, has increased in the Pacific Ocean in recent years, and because it feeds on the live corals has destroyed whole areas of coral reef.

Brittle stars

The 2,000 species of brittle stars (class Ophiuroidea), which occur at all depths of the ocean, are easily distinguished from starfish by the sharp demarcation of their central disk and by their very long arms. They also differ from starfish in several other ways. Locomotion is achieved not by tube feet but by muscular movements of the arms. Most brittle stars feed on detritus and small organisms but do not have an intestine or anus, and indigestible fragments are ejected through the mouth. Starfish (and sea urchins) breathe by

The water vascular system of echinoderms is an arrangement of fluid-filled canals. A porous structure (madreporite) links the aboral surface with a ring canal near the oral surface. Radial canals run from this water ring along the underside of each arm and give rise to lateral canals which end in a bulbous ampulla; this itself ends in one or more tube feet (podia). When the ampulla contracts fluid is pushed into the podia and elongates them. As it touches a surface the center of the tip of the podium contracts to form a vacuum and sticks to the surface.

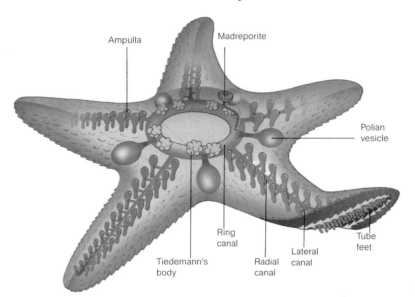

Ampulla

Madreporite

Polian vesicle

Tiedemann's body

Ring canal

Radial canal

Lateral canal

Tube feet

The brittle star *Ophiothrix fragilis* lives in dense communities on muddy, gravel substrates. It is a small animal, its disk being only about half an inch to just over an inch across. Brittle stars are the most mobile echinoderms. They move by lifting the disk off the substrate, extending one or two arms forward and trailing two behind, while the lateral arms push against the bottom with the aid of spines. This species has hooked spines on the oral surface which provide additional traction during locomotion.

means of skin gills, which are located all over the body surface, whereas brittle stars use respiratory pouches (bursae) which occur near the arm bases.

Sea urchins

Most sea urchins (class Echinoidea) have a globular body, with the skeletal plates fused together to form a hard shell (the test). In most species the body is further protected by sharp spines which may be poisonous. Some of the spines are movable, articulating in sockets in the skeletal plates, and are used for locomotion. The spines of some species of urchin are also used for boring holes in coral or rock into which the animals wedge themselves to prevent removal. In addition to spines, the skin surface of urchins and asteroids carries tiny pincer-like structures on stalks, called pedicellariae, which are used for defense, catching small prey and cleaning the body surface.

The tube feet of echinoids protrude through holes in the skeletal plates in ten longitudinal rows. In heart urchins and sand dollars *(Clypeaster* sp.) the podia are modified for respiration, and locomotion is achieved by means of the spines only. These urchins also burrow in sand, which has resulted in the spines being modifed: the heart urchin has reduced spines to assist burrowing, giving it a furry appearance, and the sand dollar has lost its spines entirely.

Sea urchins are omnivorous and often scavenge on organic debris. The mouth is on the underside (oral surface) and the anus on the upper side (aboral surface).

Sea cucumbers, sea lilies and feather stars

As their name suggests, sea cucumbers (class Holothuroidea) are sausage-shaped echinoderms, elongated along the oral-aboral axis. They have soft bodies, covered with a glandular skin, and the skeletal plates have been reduced to microscopic ossicles.

Sea cucumbers burrow in sand. To breathe, they have tubes known as "respiratory trees" which carry seawater into the body from the

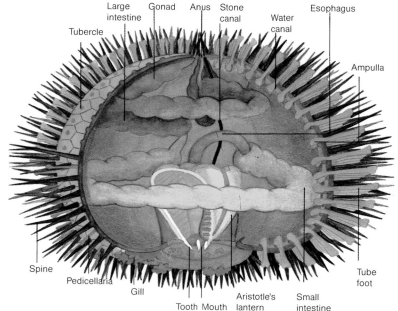

anus, so that gaseous exchange can occur internally. The podia and skin surface are also used for respiration.

Sea lilies and feather stars (class Crinoidea) are among the most primitive of echinoderms. They resemble a starfish that has been inverted and (in the case of the sea lily) attached to the sea bed by a stalk. Most fossil echinoderms are of this form. The sea lily skeleton, which stands on the stalk, is cup-shaped with arms. Tube feet are present in a groove on the upper surface of the arms, but they are used solely for respiration. Food (usually plankton) is collected by the arms and directed to the mouth on the upper surface by cilia.

The feather stars resemble sea lilies and, like them, develop from the larval stage into an attached form. They differ from sea lilies, however, in that they are free-living as adults, either creeping along the sea bed or swimming.

Sea urchins display the characteristic five-rayed symmetry most interestingly in the mouth. This consists of five pyramidal plates worked together by muscular action to scrape up food, and is called an Aristotle's lantern because it resembles an old-fashioned lamp. From the mouth the esophagus ascends and joins the small intestine which does a complete circuit inside the urchin, but when it joins the large intestine it does another circuit in the opposite direction before it joins the rectum.

The sea cucumber *(Cucumaria saxicola)* ranges from 4 to 12 inches (10 to 30 centimeters) length. Its tube feet are in arranged more or less along the pentamerous divisions. In the region of the mouth the tube feet have been modified to form tentacles which are used for food-gathering. This species feeds on plankton which it sifts through the feathery tentacles.

Mollusks

More than 100,000 species of the phylum Mollusca are known to exist, belonging to six classes. The three largest of these are the Gastropoda, which consist of snails, slugs and limpets, Bivalvia, which include oysters, mussels and clams, and Cephalopoda, containing squids, cuttlefish and octopuses. About 100,000 fossil species of mollusk have also been described, including groups such as the ammonites and belemnites (both class Cephalopoda) which are now extinct. Mollusk fossils have been found in rocks about 550 million years old and are often well-preserved because of their hard shells. But despite their age and the intact fossils, the origin of mollusks and their relationship to other invertebrates is still obscure.

Mollusk anatomy
The three main mollusk classes look very different, but all follow the same basic plan. The main bulk of the body contains the internal organs and is called the visceral mass. This mass lies above a muscular foot, and is surrounded by the mantle — a fleshy extension of the body wall — which hangs down on each side of the visceral mass. The space between the mantle and the visceral mass is called the mantle cavity. The mantle secretes the shell, which is present in most mollusks.

Almost all aquatic mollusks breathe with the aid of ctenidia (comb-like gills in the mantle cavity) which are covered with many small, hairlike structures called cilia, whose rhythmic movement draws water over the gill surface. Oxygen is taken up from the water current by blood which flows through the gill filaments, and carbon dioxide diffuses out of the gill filaments into the exhalant current.

The kidneys and anus open into the mantle cavity and waste is also carried away by the exhalant current. In some species, the edges of the mantle may be joined to form tubes called siphons, which create a one-way flow of water through the mantle cavity. Many mollusks also have sensory cells on the edge of the membrane of each afferent gill. These organs (called osphradia) are thought to be chemoreceptors, which detect the juices of prey or other food sources. They also determine the level of sediment in the incoming water, too much of which would block the delicate gill filaments.

Gastropods
The ancestral gastropod had a mouth and anus at opposite ends of the body, as do the larvae of present-day gastropods. But early in the development of larval gastropods a remarkable change takes place — the visceral mass twists through 180°. This process, known as torsion, is important because it brings the mantle cavity from a lower posterior to an upper frontal position. The shell then only needs one opening, and the mantle cavity provides a space into which the young gastropod can withdraw for protection.

The problem of fouling by waste released over the head is avoided by having holes in the shell which direct excrement away from the head (as is found in keyhole limpets and the abalone), or by shifting the anus to the side. A one-way water flow through the mantle cavity also avoids this problem.

The left and right sides of the gastropod body grow at different rates because of the spiral shape of the shell — the organs on one side do not develop, and the visceral mass, mantle and shell become spirally coiled; this development also makes the long digestive system more compact.

Most aquatic gastropods breathe by means of gills, but the land-dwelling snails and slugs (subclass Pulmonata) lack gills, and have modified the mantle cavity into a lung. Air is moved in and out of the mantle cavity by raising and lowering the mantle, which is moist and has a rich blood supply — essential features for gaseous exchange to take place.

Pulmonates show another adaptation for a terrestrial existence in that nitrogenous waste is excreted as uric acid crystals. This is in contrast to

The common garden snail, *Helix* sp., is typical of terrestrial pulmonates and shows clearly how the gills that lie in the mantle cavity of aquatic gastropods have evolved into a lung in terrestrial gastropods. The visceral mass becomes twisted during torsion, but in addition, the anal and pulmonary openings come to lie next to each other at a single opening in the shell, the pneumostome. The position of the pneumostome at the side of and behind the head helps to prevent fouling.

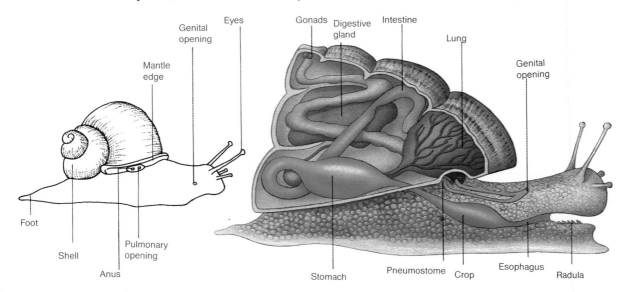

Foot · Shell · Anus · Pulmonary opening · Mantle edge · Genital opening · Eyes · Gonads · Digestive gland · Intestine · Lung · Genital opening · Stomach · Pneumostome · Crop · Esophagus · Radula

aquatic gastropods which excrete nitrogenous waste as ammonia, which is toxic in concentrated form and therefore requires a large amount of water to dilute it.

Gastropods are generally considered to be inhabitants of damp places, but the terrestrial adaptations and, more importantly, the possession of a shell into which the animal can withdraw and seal itself off from the environment have enabled them to invade such unlikely regions as deserts. Some pulmonates have secondarily returned to an aquatic existence but they still breathe by means of a lung.

Most gastropods move on a flat sole which in many species has a large pedal gland. This gland secretes mucus onto the surface over which the sole moves. Locomotion is achieved by waves of muscular contractions which pass down the foot. In the sea hare *(Aplysia* sp.) the foot is formed into folds, or parapodia, which may be used for swimming, and in the sea butterflies (Pteropoda) the parapodia are drawn out into membranous wings.

Most gastropods feed using a radula — a serrated band of teeth — to rasp off fine particles from their food. Many species, such as land snails and limpets, feed on vegetation; others, such as whelks, are carnivorous. They detect their prey by means of the osphradium. The mouth of carnivorous gastropods often lies on an extension of the head, called the proboscis. The radula of these species has fewer, larger teeth than the plant-feeding gastropods; in the cone shells *(Conus* spp.) it has become a stalk with which they stab and inject a poison into their prey. Carnivorous gastropods damage oysters by drilling through part of the shell using the radula. This is achieved by secreting a chemical which softens the oyster shell, which is then worn away by the radula. By alternately softening and rasping the shell the carnivore makes a hole in the shell through which it can insert the proboscis to eat the oyster.

Aquatic gastropods have separate sexes, but snails and slugs are hermaphroditic. The sperm of dioecious gastropods are either shed into the

water where they fertilize eggs from the females, or are introduced into the female with a penis. Eggs are deposited singly, in strings, or in a thin shell. Some are carried in the female until they hatch.

In some gastropods, the young pass through two stages: the first is the "trochophore" stage in which the larva is roughly spherical with a band of cilia for movement. It later develops a shell to become the veliger larva. In other gastropods the trochophore stage remains in the egg which hatches to produce the veliger, and in some species both the trochophore and veliger stages remain in the egg.

Some aquatic snails transmit human platyhel-

Sea slugs are gastropods which, in secondary adaptation, have become untwisted so that the mouth and anus are at opposite ends of the body. The cluster of appendages on this species are secondary gills arranged around the anus. Some sea slugs are protected by the stinging cells of their jellyfish prey, which settle in the extensions of the sea slug's body.

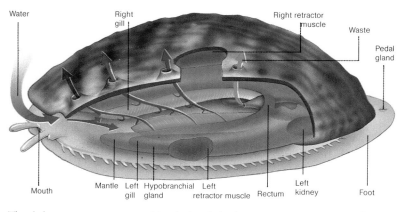

Water | Right gill | Right retractor muscle | Waste | Pedal gland
Mouth | Mantle | Left gill | Hypobranchial gland | Left retractor muscle | Rectum | Left kidney | Foot

The abalone, or ormer *(Haliotis tuberculata),* is a primitive gastropod which lives in shallow water. Its broad shell is asymmetrical with a single spiral only and, in this species, has five holes. The incoming water bathes the gills and gas is exchanged. The exhalant current removes carbon dioxide and waste and flows out through the holes.

occur on the edges of the mantle, and these detect light and test inhalant water currents.

Most bivalves are filter-feeders, filtering very small food particles from the water. The gills are highly modified for this purpose and secrete mucus in which food particles are trapped. A complex arrangement of cilia draws water into the mantle cavity, sorts the particles and carries food to the mouth. Fleshy folds near the mouth (palps) sort the trapped particles, and non-food material is passed out with the exhalant current while food particles pass into the stomach. There a gelatinous rod called the crystalline style is rotated by cilia and rubs against the stomach wall and style sac, releasing enzymes which partly digest the food particles. Further digestion occurs within the cells of the digestive glands that surround the stomach.

Most bivalves use the foot as a burrowing organ. The two valves close and water is pushed out of the mantle cavity which loosens the mud and makes the movement of the foot easier. (Some species have serrated shell edges which also break up the mud and facilitate burrowing.) Blood is then pumped into irregular channels, called blood sinuses, in the tissue of the foot. This action makes the top of the foot swell and anchors the animal. Contractions of the foot muscles then pull the animal down.

Most bivalves are sluggish, but some such as the razorshell clam have a thin, streamlined shell and large foot and can burrow very rapidly. Scallops use jet propulsion to swim, by clapping the two valves of the shell together. Other species, such as oysters (which cement themselves to rocks) and mussels (which attach themselves to surfaces by strong threads of organic material to prevent being dislodged by waves) are unable to move.

In most bivalves, males and females shed sperm and eggs into the water where fertilization pro-

The Queen scallop (*Chlamys opercularis*) is a bivalve mollusk which is mainly sessile. It swims by jet propulsion, clapping its valves together, but usually only when disturbed. During the evolution of mollusks these animals lost their head, with the result that all their sense organs are found on the edge of the mantle. These organs consist of eyes and tentacles which carry tactile and chemoreceptor cells. Scallops filter-feed, and breathe through W-shaped ciliated gills which direct a complex pattern of water currents through the body.

minth parasites such as the liver fluke and the blood fluke (*Schistosoma sp.*), the cause of bilharzia.

Bivalves

The class Bivalvia includes clams, oysters and mussels, and is characterized by the possession of a shell which is divided into two halves joined by an elastic ligament.

Bivalves are mainly marine although some have invaded freshwater. The foot is adapted for burrowing mainly in mud and sand, but some species can bore into hard material; the shipworm (*Teredo navalis*), for example, drills into wood using the roughened edges of its shell, and other species even bore into rocks.

The body is long and laterally flattened, surrounded on each side by two lobes of the mantle. Each lobe secretes a shell called a valve. The gills hang from the roof of the mantle cavity and lie on each side of the body. They are used for collecting food as well as for breathing. Bivalves have no head — it would be impossible to perceive stimuli while buried in the sand — so the sense organs

The squid (*Loligo* sp.) is carnivorous, feeding mainly on fish and crustaceans. It locates its prey with its highly-evolved eyes, darts towards it by rapidly ejecting water from the mantle cavity and seizes the prey with its two tentacles. The prey is pulled to the mouth and held by the squid's arms while its head is bitten off by the predator's powerful horny jaws. Large chunks are torn off the prey, pulled back by the radula and swallowed. Only the gut and tail are rejected.

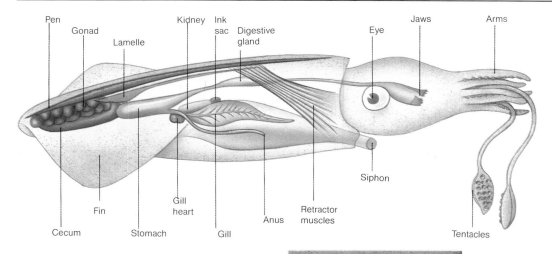

Pen · Gonad · Lamelle · Kidney · Ink sac · Digestive gland · Eye · Jaws · Arms

Cecum · Fin · Stomach · Gill heart · Gill · Anus · Retractor muscles · Siphon · Tentacles

Squids, like all cephalopods, are highly adapted to a deep-sea, carnivorous existence. The water that flows through them is used for locomotion and as the animals' oxygen source. They can cruise or hover as well as dart, using the fins as stabilizers. They have eight arms and two tentacles to catch prey with. Food is absorbed not in the digestive gland as in other mollusks but in the cecum.

duces a trochophore larva, which develops two valves to form the veliger larva. The veliger swims around in the open sea for several months before settling on the bottom to become an adult. In the freshwater mussel, however, the female carries eggs on her gills. Sperm enter the mantle cavity with the water currents and fertilize the eggs there. The larvae then leave the mantle cavity and live parasitically on the gills of fish, until they develop into adults.

Cephalopods

The cephalopods represent the highest stage of molluskan evolution. The three main groups belong to the subclass Coleoidea and are the squids (order Teuthoidea), the cuttlefish (order Sepioidea) and the octopus (order Octopoda).

All cephalopods are adapted to a free-swimming life in the sea, but only one living species has an outer shell — the nautilus. Most other cephalopods have an inner chambered shell, which varies from species to species. In the cuttlefish it is called a cuttlebone and in the squid, a pen. It lies along the dorsal parts of the animal's body and is mostly filled with gas. Toward the tail end, however, it is filled with liquid, which adds weight to that part, whereas the rest of the air-filled structure provides lift in water.

In the cuttlefish and the squid the shell has been reduced to a small plate and their bodies are elongated and streamlined. The octopus, however, is less streamlined and lacks a shell. The edges of its mantle are fused together forming a bag around the body, from which the head protrudes, surrounded by tentacles. The octopus usually crawls around the sea bed using the suckers on its tentacles, but is capable of jet propulsion when fast movement is necessary. The propelling movement is achieved by pumping water into the mantle cavity and forcing it out through a funnel below the head.

All cephalopods are carnivorous. Cuttlefish and squid catch their prey with two long tentacles with suckers at the tips, and hold it near the mouth with the eight shorter arms, which are equipped with suckers all along their length. The octopus, however, has eight tentacles of equal length, any one of which is used to grab the prey.

Retina · Ciliary muscle · Lens · Cornea · Iris · Optic nerves

The cephalopod eye is highly developed and resembles the vertebrate eye in structure. The retina contains photoreceptors which are pointed toward the light source, whereas in vertebrates the cones point away from it. An image is formed, but how clear it is is not known. Color vision is also present.

The prey is bitten with the two horny jaws and the octopus' poisonous saliva then enters the wound and kills the prey. Large pieces of flesh are torn off, pulled into the mouth by the radula and swallowed.

Cephalopods have the most highly developed nervous system of all mollusks. The brain is large and, at least in the squid and octopus, can learn and remember information. Most cephalopods rely on their good eyesight to detect food and enemies. In addition, sensory cells on the octopus' tentacles detect minute concentrations of chemicals which also alert it to the presence of prey or to possible danger.

Many cephalopods can camouflage themselves in seconds, changing color by the expansion or contraction of small bags of pigment in the skin. Some species also have an ink sac — a pouch off the intestine — which releases a cloud of black liquid through the anus when the animal is alarmed, behind which it can make its escape. It is the color of this ink that gives some of these animals their Latin name — Sepia.

Male cephalopods produce sperm enclosed in a case called the spermatophore. One of the male's arms is modified to transfer the spermatophore into the mantle cavity of the female, where the eggs are fertilized. The female may stick them onto rocks, where they develop into adults without passing through a trochophore or veliger stage.

Introduction to arthropods

The phylum Arthropoda contains almost a million species of free-living, parasitic or sedentary animals, and includes crustaceans, spiders and insects, as well as many smaller groups. The phylum is thought to be descended from primitive annelids (probably the polychaetes) because they share some characteristics, such as a metamerically segmented body — segmented along a central axis — and appendages on each segment. They also have a similarly structured nervous system. But other arthropod features represent a more sophisticated evolutionary development.

An arthropod's exoskeleton, which is jointed to allow movement, is hardened and does not grow with the animal. As a result, it must be periodically shed — a process known as ecdysis — in order to allow growth. The arthropod body is usually divided into head, thorax and abdomen, although in some groups there may be no clear distinction between the three regions. The head and thorax may be fused, forming a cephalothorax, or the abdomen may be reduced in size. The head typically carries feeding and sensory appendages. Each segment of the body usually has a pair of jointed appendages, which are modified for specific functions in different species.

Arthropods have a body cavity (coelom), but it is small and contains only the gonads and excretory organs. The other internal cavities form a hemocoele (blood cavity). The circulatory system is open, with a dorsally situated heart. The digestive system consists of a tubular gut which runs from the mouth to the anus. The foregut and hindgut are lined with chitin and are shed at ecdysis. Excretion is through specialized coelomoducts and the anus is terminal.

The nervous system has a dorsally positioned brain and a ventral nerve cord which has branches called ganglia in each segment. Tactile bristles are also a common feature. Eyes, which may be simple or compound, are usually present — compound eyes are unique to arthopods and allow the formation of a highly-defined image.

Trilobites

The trilobites are a fossil group of marine arthropods (class Trilobita) which were once extremely numerous but became extinct toward the end of the Paleozoic era, about 230 million years ago.

Trilobites measured from 0.2 inch to 3 feet (0.5 centimeter to 1 meter) long, and the body was divided into a head (cephalon), trunk and tail (pygidium). The cephalon was composed of four or five fused segments and carried a pair of antennae, a mouth, a pair of eyes and four pairs of biramous (forked) appendages, which functioned as both gills and legs. The segments of the trunk were articulated whereas those of the pygidium were fused to form a solid shield. Each segment possessed a pair of walking legs.

Horseshoe crabs

Despite their name, horseshoe crabs (subclass Xiphosura) are not crabs at all, but primitive marine arthropods. They are the only living group of the class Merostomata, of which only five species remain.

The body is divided into three main regions: the horseshoe-shaped prosoma (fused head and thorax), covered by a thick shell, or carapace, and ending in a long, pointed tail called a telson. On the ventral surface of the prosoma is a mouth, with a pair of pincerlike feeding appendages called chelicerae, and five pairs of walking legs. There are five pairs of abdominal appendages which are modified as gill-books, which are similar to the lung-books of spiders; they keep a con-

The dorsal exoskeleton of trilobites was much thicker than the ventral one, which is why most fossils display the dorsal view. The name Trilobita derives from the triple-segmented transverse divisions that run down their length. They were bottom dwellers and scavenged for food.

This species of horseshoe crab *(Limulus polyphemus)* is found in the shallow coastal waters of Asia, the Gulf of Mexico, and the northern Atlantic. Its telson, which it uses to push with and to right itself, is also used to make threatening gestures, as in this position.

stant current of water circulating over the gills, and also act as paddles, providing propulsive power for the animal. The nervous system is well-developed, but the eyes are not; they can detect movement, although they cannot form an image.

During mating, horseshoe crabs gather together in huge numbers at the water's edge. Eggs are laid in the sand, fertilized by the males and then covered. Newly-hatched horseshoe crabs are called trilobite larvae because of their similarity to the larvae of that class.

Pycnogonids

The class Pycnogonida contains about 600 species of carnivorous marine animals known as sea spiders. They feed on corals and sponges, and are found in all marine waters. The head bears chelicerae, a mouth which is a cylindrical proboscis, and four eyes. The similarity of pycnogonids to true spiders is only superficial. Pycnogonids have a segmented abdomen and legs which end in claws, whereas spiders do not have these characteristics.

Chilopods and diplopods

The animals in the class Chilopoda are commonly known as centipedes and those in class Diplopoda, as millipedes.

Centipedes have one pair of antennae on the head, a pair of mandibles and two pairs of maxillae, in contrast to millipedes which have at least the front pair of maxillae modified to form a lower lip (gnathochilarium). Centipedes have eyes that may be simple ocelli (a cluster of photoreceptors) or modified compound eyes, whereas if eyes are present at all in millipedes, they are only simple ocelli. Centipedes are flattened and divided into a large number of segments, each of which, apart from the first, carries a pair of long, slender, walking legs.

Millipedes are vegetarian scavengers. Their body is also segmented, but is usually cylindrical. The first four trunk segments differ from the others in that the first is legless and the following three bear only one pair of walking legs. These four segments together are sometimes called the thorax. The abdomen has many segments which are fused together in pairs (diplosegments). Each diplosegment bears two pairs of short legs, and as a result a millipede may have as many as 115 pairs of legs. Nevertheless, they are not good runners and, when attacked, they protect themselves by coiling up, and many species emit a foul-smelling secretion from special stink glands on the sides of the body.

Millipedes and centipedes are found mainly in damp conditions such as rotting logs or in leaf litter, because they do not possess a waxy cuticle with which to reduce water loss.

Both groups have separate sexes, and the female lays eggs which are fertilized by the male. Some species lay in the eggs in a "nest" where they are guarded by the female, but others, such as the centipede *Lithobius,* lay one egg at a time and then leave it. The young resemble the adult and in

some species of centipedes have the same number of segments, but other young centipedes and all millipedes have fewer body segments than the adults do.

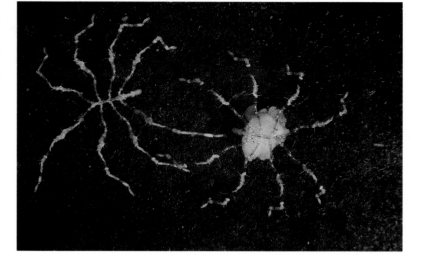

Pycnogonids, or sea spiders (above), such as this species *Endeis pauciporosa,* are notable for their ovigerous legs. In the male they are well developed and are used to carry around masses of fertilized eggs until they hatch.

Centipedes (below) are carnivorous and kill their prey with a large pair of poison claws (maxillipeds) found in the first body segment. Each claw has a pointed fang which is fed by a poison gland. This cave centipede *(Scolopendra* sp.) is eating a cricket.

Crustaceans

The crustaceans became the major group of marine arthropods when the trilobites and giant sea scorpions (Eurypterida) became extinct. The size of the group has tempted some zoologists to make it a subphylum — it contains about 30,000 species and includes such animals as water fleas, barnacles, crabs, lobsters, shrimps and wood lice (pill bugs). It also includes tiny zooplankton, which live near the surface of the sea and occupy an important position in aquatic foodchains.

Characteristic features
The exoskeleton, which is composed of chitin and in some species (such as crabs) is hardened with calcium salts, is molted periodically to enable the animal to grow. This is an important process in the life of crustaceans (and other arthropods) and occurs in a strict sequence of events. First, useful materials such as calcium salts are reabsorbed into the body; the new cuticle is then laid down underneath the old one; the animal swells up causing the old cuticle to split, and finally the new cuticle is hardened. Before the new cuticle hardens the animal is easy prey for predators and usually seeks shelter during this period.

In many crustaceans an outer shell, or carapace, covers the exoskeleton of the thorax or anterior trunk segments. Typically the body consists of a head, thorax and an abdomen, but often the head and a number of thoracic segments are fused to form a cephalothorax. The head region has two pairs of antennae, which is characteristic of crustaceans. It also has one pair of mandibles, which in most species are heavy and have grinding and biting surfaces, and two pairs of maxillae.

The trunk region of the cephalothorax is segmented, the number of segments varying according to the species, and each segment bears a pair of appendages. These appendages are different in each species, being modified to suit particular functions. Each abdominal segment usually also has a pair of appendages structured for various functions such as swimming, crawling, respiration, food capture or reproduction.

The appendages are biramous — that is, each one ends in two jointed branches; they are tubular and jointed, and contain muscles which contract to bend the limbs. Crustaceans swim by beating these appendages — some of which have a fringe of bristly setae, to push against the water — and most of the animals also crawl.

Many crustaceans breathe with gills, although some terrestrial species, such as the robber crab *(Birgus latro)*, have modified them to become air chambers lined with blood vessels which absorb oxygen. The gills are also a secondary excretory organ — the primary ones being the antennal and maxillary glands in the segments bearing those structures. The crustacean blood system is open — the blood is pumped by the heart into a hemocoele where it simply bathes the tissues.

The nervous system of these animals is well developed, and the sensory organs include eyes and various tactile receptors. Most crustaceans have compound eyes in the adult stage. The young, or nauplius larvae, have a median eye composed of three or four ocelli (clusters of photoreceptors), which in some species persist into the adult stage. Other sensory receptors include special tactile hairs, which are scattered over the body surface but are most concentrated on the appendages, and statocysts (balancing organs).

Crustaceans are usually dioecious — that is they have separate sexes — but some are hermaphroditic and have both male and female sexual organs. The young hatch from eggs which are usually brooded and take the form of free-swimming planktonic larvae with fewer appendages than the adult, but through successive molts, trunk segments and extra appendages are acquired.

Branchiopods
This subclass of mostly freshwater animals have earned their name (meaning "gill feet") from their thoracic appendages which are modified for respiration as well as filter-feeding and locomotion. The group consists of four orders: Anacostraca (fairy shrimps), which have no carapace; Notostraca (tadpole shrimps), whose head and front thoracic region is covered by a carapace; Conchostraca (clamp shrimps) and Cladocera (water fleas). The animals in this last group, which include the common genus *Daphnia,* are covered by a carapace which encloses the trunk but not the head. The carapace of these animals is usually transparent, but can appear red or pink, depending on the level of oxygen in the water. *Daphnia,* for example, is transparent in oxygenated water, but turns pink in stagnant water, when it produces hemoglobin as a respiratory pigment.

Daphnia, a genus of water flea, is a typical branchiopod. It has five or six pairs of trunk appendages which have epipodites — extensions that serve as gills. The appendages also have setae on them which collect food particles. The food is then moved into a food groove and spurts of water push it toward the mouth. Unlike other branchiopods, which use their trunk appendages for locomotion, *Daphnia* uses its large second antennae as paddles to move it in a jerky up-and-down motion. Also, water fleas have a single median compound eye which is formed by the fusion of two compound eyes and directs them when swimming.

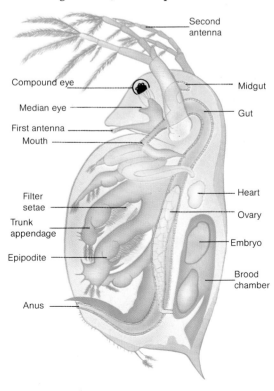

- Second antenna
- Compound eye
- Median eye
- First antenna
- Mouth
- Midgut
- Gut
- Filter setae
- Trunk appendage
- Epipodite
- Heart
- Ovary
- Embryo
- Brood chamber
- Anus

Branchiopods are filter-feeders which collect food particles on the bristles on the trunk appendages. The food is carried to the mouth by streams of water which pass between the trunk appendages as they move back and forth. The first maxillae finally push the food into the mouth.

The sexes are generally separate and copulation is the means of fertilization, but under favorable conditions the females reproduce parthenogenetically (when the eggs develop without fertilization). The eggs of water fleas hatch into females for several generations until adverse conditions (such as low water temperature, or short food supply) induce the production of males, after which eggs are again fertilized by copulation.

Ostracods and copepods

These two subclasses of tiny crustaceans contain freshwater and marine species, and planktonic (shallow-water) as well as benthic (deepwater) types. The ostracods are similar to the conchostracan branchiopods in that they have a bivalve hinged carapace. The head is more developed than

the remainder of the body, and its appendages (especially the antennules and antennae) are modified for crawling, swimming and feeding. Most ostracods are filter-feeders, but some species are scavengers, predators or parasites.

Most copepods are marine and occur in huge numbers, making them of great economic and ecological importance, because they form a large part of the diet of many fish. The body is short and tubular, with a trunk that is usually composed of six segments, an abdomen and — in contrast to the ostracods — a reduced head region. The first antennae are long, whereas the second pair is short and often branched. Rapid locomotion is achieved by the beating of the thoracic appendages which causes jerky movements, but a slower gliding motion results from the movement of the second antennae.

The planktonic copepods usually live in the upper 650 to 975 feet (200 to 300 meters) of the ocean, but diurnal vertical migrations are common. Light appears to be the trigger for this behavior but its value to the animal is far from certain.

Cyclops sp. are freshwater copepods. These animals are remarkable in that they can stagger the development of eggs that are produced from a single mating. After fertilization, one or two ovisacs form on the genital segment of the female — each sac holds up to 50 eggs. A number of eggs are hatched from half a day to five days after fertilization and a new batch is then brooded. The eggs hatch as nauplius larvae, just as the one shown above, swimming away from the parent.

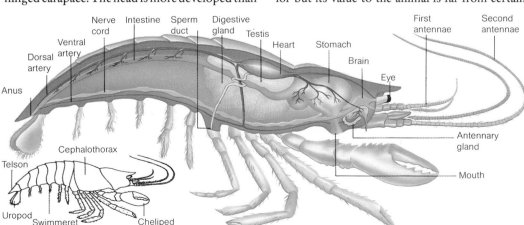

The anatomy of crayfish *(Cambarus* sp.) is similar to that of most decapods but the male (left) and female species differ in some respects. For instance, the female has large swimmerets on which she carries newly-hatched larvae, whereas those of the male are small. Instead, the male appendages are modified for the transmission of sperm.

Barnacles can be stalked, such as the goose barnacle *(Lepas* sp.) which is cemented to the substratum by a peduncle, or non-stalked and attached directly to a surface. The peduncle of the goose barnacle carries the capitulum, or body, which is surrounded by calcareous plates. The dorsal plate is called the carina.

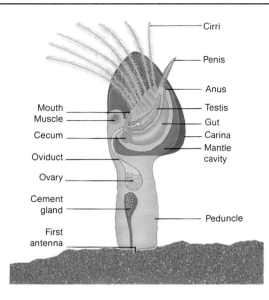

Many copepods are filter-feeders, some are predatory, others are scavengers and still others are parasitic. The parasitic species feed on freshwater and marine fish, some living on the gill filaments, and others in the intestines.

Neither ostracods nor copepods have gills, and gaseous exchange occurs in the valves (in ostracods) or across the body surface (in copepods). Both groups have an open circulatory system. Median eyes are found in both and only one group of ostracods has compound eyes. One outstanding feature of these animals is their luminescence. Both are dioecious and fertilization is generally by copulation, although parthenogenesis is known in some ostracod species.

Barnacles

Barnacles belong to the only group of sessile crustaceans, the Cirripedia. The subclass derives its name from the six pairs of feather-like trunk appendages (cirri) that protrude from the shell, which are used for filter-feeding.

All barnacles are marine, and most are found attached to rocks, shells or driftwood; there are, however, a few commensal and even some parasitic species, which live on other crustaceans. They are surrounded by a carapace of calcareous plates, which develops from a bivalve, clamlike carapace in the young, or cypris, larva.

Most barnacle species are hermaphrodite, but cross-fertilization is common. The young hatch as nauplius larvae, as in other crustaceans, but then develop into cypris larvae which settle on a suitable surface. Finally metamorphosis takes place and through a series of molts the animal rapidly adds the carapace and reaches adulthood.

Decapods

The order Decapoda contains most of the larger crustaceans in the subclass Malacostraca; they include lobsters, crabs, crayfish and shrimps. More than 8,500 species have been described and most are marine, although some species do live in freshwater. All decapods have eight pairs of thoracic appendages; the rear five are modified for walking and are the origin of the name decapod; the front three pairs serve as feeding appendages (maxillipeds). In some decapods the first pair of walking legs is heavier and stronger than the others and has chelae (pincers).

The abdomen is composed of six segments and a tail (telson). The abdominal appendages are called swimmerets and in some species these are reduced. The sixth abdominal segment usually has a pair of appendages called uropods, which together with the telson form a tail fan.

Decapods have compound eyes on jointed movable stalks, and the central nervous system is well developed. Their wide variation in color usually depends on the habitat and the pigment-producing chromatophores in the exoskeleton. Aquatic decapods breathe with gills, usually five pairs, which run vertically in the cephalothorax between the endoskeleton and the other organs.

The shore crab *(Carcinus maenas)* is one species in a group that includes scavengers and predators of other large invertebrates, such as the lugworm, *Arenicola marina.* The prey is attacked with the chelipeds and caught up by the maxillipeds, which pass it to the other mouthparts. A portion is bitten off, the rest is torn apart and the morsels are fed into the mouth. The heavy claw on the right has blunt serrations on the pincers, which are used for crushing. The lighter claw on the left is modified for cutting.

Krill (*Euphausia* sp.) are not decapods (even though they are shrimplike in appearance) but belong to the same class (Malacostraca). Like some shrimps, some species of krill are luminescent. They are well-designed for swimming but do not have a carapace completely enclosing the gills, as do most decapods. Most krill are filter-feeders, feeding off zoo- and phytoplankton.

Water flows in at the base of the front five appendages, bathes the gills and flows out through a vent under the second antennae.

An elaborate courtship ritual is typical of many decapods prior to copulation and odorous chemicals called pheromones play an important role in their sexual behavior. In many species the eggs are laid soon after copulation and are cemented onto the swimmerets of the female.

Lobsters, like some crabs, have a single huge claw that is used for crushing; the other claw of the pair is much smaller but has sharp edges and is used for seizing and tearing prey. They are scavengers, but also catch fish and break open shelled animals. Crayfish are similar in appearance to lobsters but most live in freshwater whereas lobsters are marine. Also, like lobsters, crayfish are nocturnal and feed on almost any organic matter, living or dead.

Crabs are probably the most successful decapods in that they can live on land as well as in water. They are found at all depths of water, in all parts of the world. They have a wide carapace and, unlike lobsters, a small abdomen which is tucked tightly under the cephalothorax. They do not have uropods and the female uses its swimmerets only for brooding eggs. They can walk forward, but more usually move sideways; the large front claws (chelipeds) are not used for locomotion.

Crabs range in size from the pea crabs (Pinnotheridae), which live in the tubes constructed by marine annelids, on sea urchins, or in the mantle cavity of gastrophods, to the Japanese spider crab *(Macrocheira kaempferi)* whose body measures about one inch (2.5 centimeters) across and whose leg span is more than 3 feet (1 meter). Their diversity of form includes the hermit crab which has no shell of its own and which takes over empty gastropod shells for protection.

Crabs are filter-feeders, predators and scavengers and their method of obtaining food is usually reflected by the shape of their chelipeds.

Shrimps and prawns are much better designed for swimming than the lobster, being laterally compressed with a well-developed abdomen, but even so most of them are bottom-dwellers. Their thoracic legs are usually long and slender and the first three pairs may have claws. The abdomen has long swimmerets which are fringed and used for swimming and in females eggs are attached to them.

Isopoda

This order contains the only group of truly terrestrial crustaceans — the woodlice — although most species are marine. The shield-shaped body is flat, has no carapace, and has seven pairs of legs. They have a pair of compound eyes and two pairs of antennae, one of which is vestigial.

Water loss can be a problem to woodlice as they do not have a waxy cuticle like some other terrestrial arthropods, such as insects. They survive by living in fairly damp habitats and by leading a nocturnal existence. Other behavioral adaptations, such as rolling up into a ball, help to reduce water loss.

Wood lice are the only species of terrestrial crustaceans. They have adapted to the drier conditions by living in moist environments, such as leaf litter and under stones, and by adopting a nocturnal life style. Moisture is essential because these animals still breathe by means of gills. They feed on decomposing organic matter, although some species are carnivorous.

Arachnids

The class Arachnida is a group of mostly terrestrial arthropods which includes spiders, scorpions, mites, ticks, harvestmen, false scorpions and sun spiders. Many arachnids have highly developed chelicerae (also found in pycnogonids and horseshoe crabs) in the head region, which usually take the form of a pair of pincer-like organs. It is possible that arachnids represent a migration of chelicerates from the sea to the land. In doing so, they have developed a cuticle, which reduces water loss, and their gill-books have become lung-books.

Arachnid anatomy

Like all arthropods, arachnids have a hard, chitinous exoskeleton and jointed appendages, but unlike most of their relatives true arachnids have no antennae. Their bodies are divided into a cephalothorax and an abdomen. The cephalothorax (or prosoma) is usually unsegmented and its upper surface is covered with a carapace. The lower region is usually protected by sternal plates. The abdominal segments of most arachnids, apart from scorpions, are fused, and in ticks and mites both the cephalothorax and abdomen have fused to form a single body.

The chelicerae in the head region of the cephalothorax are used to grasp prey and for feeding. A pair of pedipalps, which are modified to perform various functions in different species, are also present along with four pairs of legs. Because arachnids have no jaws, they cannot chew, although most of them are carnivorous. They therefore have to take their food in liquid form. Digestive enzymes are secreted onto or injected into the prey, which is then sucked in.

The respiratory system of arachnids comprises specialized breathing organs, called lung-books, and a network of tracheae (tubes that carry air from the exterior to the internal organs). Some arachnids have lung-books only, others have tracheae only, and some have both. The circulatory system is usually open, that is, arteries carry blood from the heart into a series of blood spaces — the hemocoele. Blood is oxygenated as it flows past the lung-books on its way back to the heart. Some arachnids have hemocyanin — a copper-based oxygen-carrying compound — as the major respiratory pigment of the blood. The excretory system includes a pair of Malpighian tubules which open into the alimentary canal, and coxal glands, which open to the exterior in the cephalothorax.

The orange garden spider *(Argiope aurantia)* is a typical web-spinner. Like all spiders it has eight legs joined to the cephalothorax, which also bears a pair of pedipalps and fanglike chelicerae. Most spiders have eight eyes. Apart from the brain and sucking stomach (which are in the cephalothorax) most major organs are in the abdomen, including the silk glands and their spinnerets.

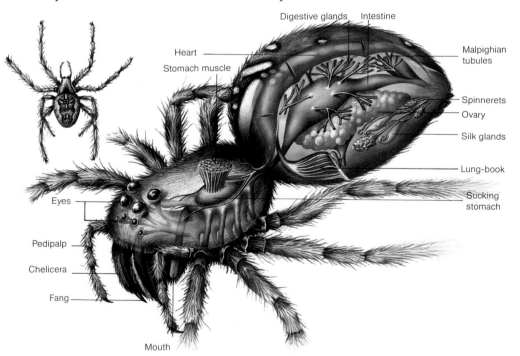

Digestive glands — Intestine
Heart
Stomach muscle
Malpighian tubules
Spinnerets
Ovary
Silk glands
Lung-book
Sucking stomach
Eyes
Pedipalp
Chelicera
Fang
Mouth

An orb web (the stages of construction are shown below right) takes a spider about an hour to make. Many species spin a new web every day, usually at night or just before dawn. This may be necessary because the web has been damaged, for instance by the struggles of trapped insects. Rain can damage the webs of some species, although others, particularly in the tropics, build webs strong enough to withstand a storm.

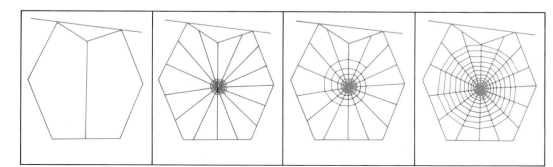

Courtship in scorpions involves a curious type of dance. The male grabs the female's pincers in his own (so that she cannot use them) and, with their tails held upright, the pair move backward and forward. After a while the male drops a packet of sperm on the ground and drags the female over it, when she takes it up into her body.

The brain consists of two ganglia (bundles of nerve cells) above the esophagus, joined to more ganglia below it. Nerves from the ganglia run to various sensory organs and there are sensory hairs over the body surface. Most arachnids have simple eyes — spiders usually have eight, scorpions twelve; even so, their eyesight is, in most cases, very poor.

Reproductive systems vary in arachnids, but the sexes are generally separate. Fertilization is by copulation, usually preceded by an elaborate courtship ritual. Sperm is often transmitted through a packet of sperm, or a spermatophore, which is picked up by the female. When hatched, most arachnids are small versions of the adults; they do not metamorphose as do many insects. The young molt several times while growing.

Spiders

There are 29,000 known species of spiders, the largest of which are the tarantulas of South America (suborder Orthognatha), some species of which have a body 3 inches (8cm) across with a leg spread of 7 inches (18cm). Tarantulas have poison glands in their chelicerae, but most spiders have them in the cephalothorax, although the poison is administered via the chelicerae. The poison is used to paralyze or kill prey, or in self-defense.

Spiders have abdominal glands which exude silk through organs called spinnerets. The silk threads are used to make webs, to catch prey and to make cocoons. Not all spiders spin webs, however; trap-door spiders (family Ctenizidae) dig a tunnel often more than 10 inches (25 centimeters) deep, which they line with silk. The tunnel is closed at ground level by a hinged door. When a small animal passes by, the spider jumps out and grabs it. Many wolf spiders (Lycosidae) and jumping spiders (family Salticidae) do not trap their food but stalk their prey and then leap on it.

Spiders perform a courtship ritual before mating. Male wolf spiders, for example, use their pedipalps as semaphores to attract a female. In many web-spinning spiders, the courting male plucks the threads of the web in a special way which the female will recognize and so not mistake him for prey. Before courtship begins the male spins a "sperm web" onto which he drops semen and fills

Ticks of the species *Ixodes hexagonus,* their bodies swollen with blood sucked from their host — a European hedgehog — remain attached to the skin of their host and feed off it continuously for several days.

a reservoir at the tip of his pedipalp with sperm. During mating he inserts the pedipalp into a special pouch in the female in which the sperm are stored. Eggs are not laid until they are ready when sperm is released over them. The fertilized eggs are then wrapped in layers of silk to form a protective cocoon, which is hidden or carried around until they hatch.

Scorpions

Scorpions (order Scorpiones) have remained virtually unchanged for about 450 million years. The largest of the true arachnids, reaching 5 to 8 inches (12 to 20 centimeters) in length, they are nocturnal and live in tropical and subtropical regions. Their pedipalps are powerful pincers, used to grasp prey. The segmented tail has a sharp sting at the tip with a poison gland, which causes paralysis and death to insects and small mammals.

Other arachnids

Mites and ticks (order Acarina) are small arachnids, usually less than one millimeters in length. Many are parasites, living on blood and tissue fluids, causing skin irritation and occasionally transmitting diseases to their hosts. Their chelicerae can pierce skin, making ticks, for example, difficult to dislodge. Harvestmen (order Opiliones) are spider-like animals with long legs and only two eyes. False scorpions (order Pseudoscorpiones), like true scorpions, have pincers (large chelate pedipalps), but no tail or sting. Sun spiders (order Solufugae), which can be up to 2 inches (5 centimeters) long, have simple elongated pedipalps that make them look like ten-legged spiders.

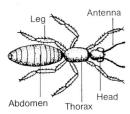

A typical insect has one pair of antennae and three pairs of legs. The body consists of three parts: head, thorax and segmented abdomen. Many insects also have wings at some stage in their lives.

Insects

No other terrestrial group of animals has as many members as the class Insecta. More than 800,000 species of insects have been described — about 300,000 of these alone are beetles (order Coleoptera) — and there are many thousands more yet to be identified and named. The success of this group is due partly to its tremendous adaptability and huge variation in life styles — insects live in almost every habitat (even down to the seashore where they are periodically inundated by the tide). The ability to fly has allowed them to spread to new, unexploited habitats and, many millions of years ago, to escape from terrestrial predators. It also helped them to disperse and has enabled greater access to food and more desirable environments.

It is possible that the great increase in the numbers and variety of flowering plants during the Cretaceous period also contributed to the enormous success of insects. Most flowering plants are dependent upon them for pollination.

General features

The exoskeleton which covers the entire insect body is composed of chitin and hardened by proteins. It is made up of several parts: the tergum, covering the back; the sternum, on the underside; and two pleura, which link the tergum to the sternum. The pleura are considerably thinner than the rest of the cuticle and generally contain spiracles — the openings to air tubes (tracheae).

The body is clearly divided into a head, thorax and abdomen. The head is derived from six segments, but they are fused together and are not obvious in the adults. Typically the insect head bears a single pair of sensory antennae, one pair of compound eyes which may be color sensitive, and one more ocelli (clusters of light-sensitive photoreceptors).

The thorax is composed of three segments, each bearing a pair of walking legs; this characteristic is unique to insects. The legs are modified in different species for grasping, swimming, jumping or digging. The winged insects (subclass Pterygota) also bear a pair of wings on the dorsal surface of each of the second and third thoracic segments, whereas those of the subclass Apterygota are primitively wingless. The abdomen is made up of eleven segments but the eleventh is often reduced. The eighth and ninth segments (also the tenth in males) bear the genital appendages.

All insects have a heart which lies dorsally within the thorax and, in most species, in the first nine abdominal segments. The blood circulates in a blood space (called the hemocoele) and bathes all the tissues.

Respiration in most insects takes place by means of a system of internal tubes called tracheae, which open to the exterior via paired spiracles. Oxygen diffusion along the trachea is sufficient to meet the demands of the insect at rest. During activity, however, air is pumped in and out of the tracheal system by expanding and collapsing air sacs (enlarged parts of the tracheae), which are controlled by movements of the body. The spiracles have closing structures for water conservation.

The principal excretory organs of insects are Malpighian tubules, which open into the hindgut and the rectum. Uric acid, dissolved salts and water are drawn from the hemocoele; the fluid in the tubules then passes into the rectum where useful salts and water are extracted before the waste products are excreted with the feces.

The insect nervous system is much like that of other arthropods, with a brain and a system of linked individual and fused ganglia connected to a ventral nerve cord. Apart from the eyes, sense organs such as chemoreceptors and tactile hairs occur all over the body, but are most concentrated on the appendages.

Insect flight

One of the most successful adaptive features of insects is flight. Most pterygotes have wings, although secondarily wingless insects do occur in

The locust (*Locusta* sp.) has the anatomy of a typical winged insect, with a head, thorax and abdomen. The thorax has three segments to which the three pairs of legs and a pair of wings are attached. A chambered heart forms part of the dorsal blood vessel, which pumps blood forward with the aid of valves. Food is taken through the mouth and is digested in the crop, the pyloric ceca, the mid- and the hindgut. The first ganglion of the ventral nerve cord forms the brain. Spiracles (not shown) open out in the cuticle of each segment. Air passes into them and carbon dioxide is released from them. The air is conducted by tracheae (ducts) to the blood, into which the oxygen diffuses.

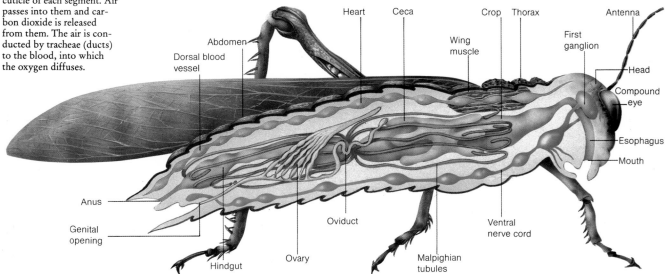

Dragonflies (A) have densely veined, primitive wings which cannot be folded across the back. But more highly evolved insects, such as wasps (B) and beetles (C), have developed structures at the wing base, called sclerites, which allow the wings to be folded across their body, and the wing venation is reduced. The two wings on either side of the wasp's body are hooked by frenal hooks. In beetles the front pair of wings have hardened and become elytra to form a wing case. They fly only with the hind wings.

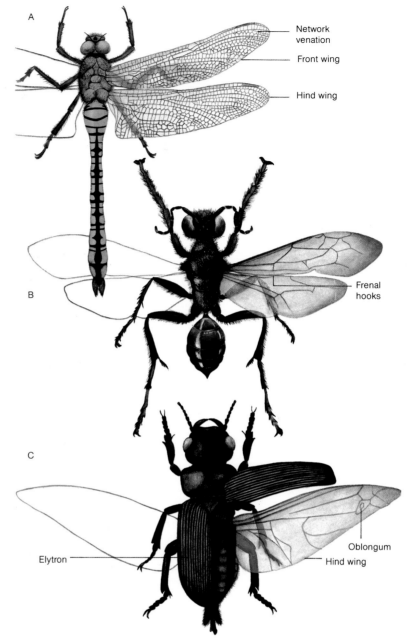

Network venation

Front wing

Hind wing

Frenal hooks

Elytron

Oblongum

Hind wing

this group. Various species, such as ants (order Hymenoptera) and termites (order Isoptera), have wings during certain stages only of their lifecycle; others have completely lost their wings, for example fleas (order Siphonaptera), as a result of their parasitic life style.

The earliest insect wing is thought to have been a fan-shaped membranous structure with heavy supporting veins. Modern wings, however, are more highly specialized structures, composed of two pieces of opposing cuticle which are separated in places by veins. These veins support the wing and provide it with blood. Primitive wings such as those of dragonflies (order Odonata) have large numbers of veins which form a net-like pattern, but subsequent evolution has favored a reduction in venation.

Many insects have two pairs of wings. These may move independently, as in damselflies (order Odonata) or they can be hooked together so that they move as a single structure, as in many hymenopterans, such as bees and wasps. Coleopterans (beetles) have undergone a further change — the first pair of wings has been hardened to form the wing cases (elytra) that protect the membranous hind wings, which are used for fight. Insects of the order Diptera, which includes all the true flies, have a pair of fore-wings only. The hind pair has been reduced and modified into club-shaped structures called halteres, which act as organs of balance.

The large compound eyes of insects are multi-faceted in those that fly and depend on sight for feeding. The eyes of some flies each consist of about 4,000 facets packed together, which are not always of equal size — those of the horsefly *(Tabanus* sp.) are larger on the front and upper parts of the eye. Each facet is formed by the cornea of a visual element called an ommatidium which contains the cone, iris and retinal cells necessary for vision.

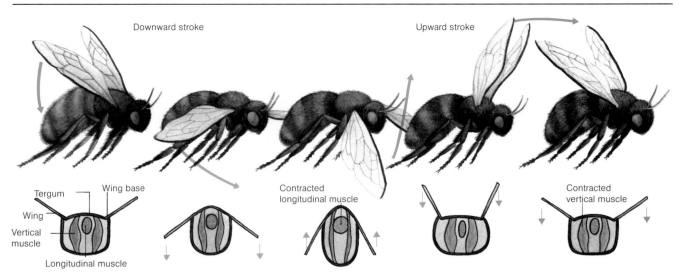

Downward stroke

Upward stroke

Tergum
Wing base
Wing
Vertical muscle
Longitudinal muscle

Contracted longitudinal muscle

Contracted vertical muscle

Insect flight involves a wing movement in the shape of a figure eight. The downward beat results from muscles at the wing base and the longitudinal muscles contracting, which also raises the tergum. An upward beat is achieved by the vertical muscles contracting.

The action of flight is fairly straightforward; an upward beat is produced by the contraction of vertical muscles in the thorax, which pulls down the tergum. A downward stroke is produced either by the contraction of muscles attached to the wing base, or by the contraction of horizontal muscles in the thorax which push the tergum up. This up-and-down movement on its own does not provide enough impetus for flight, however, and the insect's wings must also move back and forth, resulting in a wingbeat that forms a figure-of-eight or an ellipse, which is tilted at an angle to the vertical.

The number of wingbeats per second varies greatly from between 4 and 20 for butterflies (order Lepidoptera) to 190 beats per second in bees, and up to 1,000 per second in a small fly. The mode of flying also varies in different groups. Butterflies tend to have a slow, fluttering style, whereas bees and flies can hover and dart. Flight is controlled by a complex interaction of feedback from sensory hairs on the head, stretch receptors at the base of the wing, visual cues and the wing

muscles themselves. There is no nervous centers for the control of flight. Flight speed appears to be controlled by the flow of air against the antennae or by sight.

Life cycles and development

Most insects lay eggs. Once hatched the primitive wingless insects do not metamorphose — they gradually mature through a series of molts. The winged insects, however, do metamorphose. They are divided into two groups: the Endopterygota, such as flies and butterflies, whose larvae do not resemble the adult and whose wing development is internal (the derivation of their name), appearing only in the final stage of metamorphosis; and the Exopterygota, such as cockroaches, grasshoppers and some bugs, which hatch as miniature versions of the adults, having external wing buds but without being sexually mature, and gradually develop by incomplete metamorphosis (hemimetamorphosis).

The basic life style of insects can be modified by various factors; aphids, for example, exhibit parthenogenesis in favorable weather conditions. Unfertilized eggs laid in the autumn hatch into wingless ovoviviparous females, which do not lay eggs but give birth to broods of similar females. The cycle continues until conditions deteriorate, when one generation produces winged females and males, which mate.

Many insects have complicated life cycles — especially parasitic insects. Parasitic wasps, for example, have a form of reproduction called polyembryony, in which the embryonic cells give rise to more than one embryo. This means that from one egg deposited in the body of a host many larvae can be formed, and the resulting embryos can use the host's body as both a refuge and a food source.

A great variety of parasitic insects exist, from blood-sucking ectoparasites that live permanently on a host during their adult lives to the parasites that visit a host only to feed. Some, such as mud-digger wasps, are parasitic in the larval stage only. The egg-laden female wasp finds a spider which it paralyzes with its sting; it then builds a

A honey bee colony consists of workers, drones and a queen, each with distinct roles. The queen controls the colony, and lays up to 2,000 eggs a day; drones mate with the queen; and workers forage, build combs and attend to all other aspects of the colony's life.

	Description	Number in average colony	Average lifespan
	Worker Non-reproductive female	60,000 (approximately)	2 — 6 months
	Drone Reproductive male	200 (approximately)	2 — 4 months
	Queen Reproductive female	1	up to about 6 years

nest of mud into which it puts the spider. Finally, it lays an egg and seals up the nest. When the egg hatches the larva has a ready source of food until it pupates.

Feeding adaptations

The mouthparts of insects show great variability. In some they are adapted for sucking, in others for piercing and biting, and in still others the mouthparts are modified for chewing, crushing and tearing.

Chewing mouthparts are found in carnivorous and herbivorous insects. This is regarded as a primitive condition and is characteristic of insects such as dragonflies, grasshoppers and beetles, in both the adult and nymphal stages. Butterflies have chewing mouthparts during the larval stage only.

Bees and wasps have mouthparts that are modified for biting and sucking as well as chewing. In bees the sucking is done by the highly modified maxillae and labium, whereas the mandibles and labrum have retained their chewing ability, which is necessary for the manipulation of pollen and wax.

The piercing mouthparts of insects that feed on plant juices, such as aphids, penetrate plant tissues and the insect feeds on the sap. Other insects, such as houseflies, have mouthparts which are modified to form a proboscis that is used to sponge up liquid food partially digested by enzymes in the saliva.

Social organization

Communal living is developed to the highest degree in termites, ants, bees and some wasps. The social life of honey bees (Apis mellifera), for example, is controlled by a rigid division of labor and different castes of insects perform different functions within the colony. The nest is made up of a series of vertical wax combs, with hexagonal cells on both sides. Drone cells are larger than

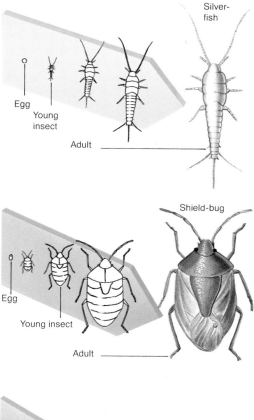

Silver-fish

Egg
Young insect
Adult

Shield-bug

Egg
Young insect
Adult

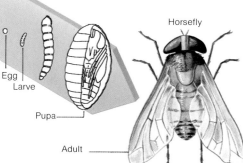

Horsefly

Egg
Larve
Pupa
Adult

Ametabolous insects, such as the silver-fish (Lepisma saccharina), develop from the egg stage into a small version of the adult. They develop without metamorphosing, in contrast to hemimetabolous insects such as the shield-bug (Chlorochroa ligata), which develops by incomplete metamorphosis into the adult. This insect emerges from an egg, which is shaped like a barrel with an operculum, or lid. The young insects resemble the adult but are wingless up to the sixth and final molt when the wing rudiments develop. The horsefly (Tabanus sp.) is an example of complete, or holometabolous, metamorphosis. Its development occurs in three stages. The first is the feeding stage, when the hatched larva eats continually and increases in size. When it stops feeding it becomes inactive and constructs a hardened cuticle, or pupa. Inside the pupal case adult structures develop at this stage from embryonic rudiments. When the period of pupation is over the adult insect emerges.

Parasitism among insects is illustrated (left) by the fate of a large white butterfly caterpillar (family Pieridae). A braconid wasp (family Braconidae) has laid eggs in the caterpillar so that emerging larvae can feed off their host.

The peacock butterfly (Inachis io, right) uses both camouflage and warning coloration for protection. When its wings are closed their dull undersides are good camouflage. But when disturbed, its wings open to reveal prominent "eye" markings, which frighten a predator.

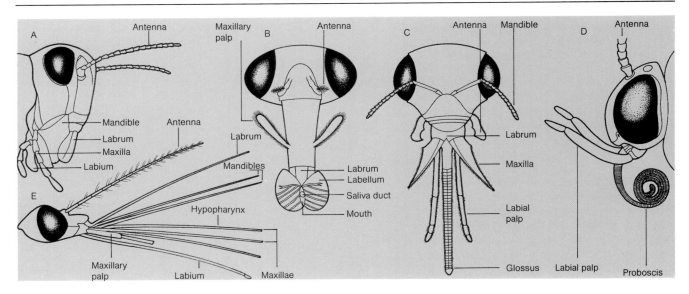

those in which workers develop, and the queen cells are more saclike. Some cells are constructed for storing pollen and honey for use during bad weather or over winter.

A colony consists typically of about 60,000 workers (all sterile females), 200 male drones and a single queen, whose function is to lay eggs. The main purpose of the drones is to mate with the queen. The duties of the worker bees alter as they become older. Newly emerged workers stay in the hive and clean and feed the queen, drones and larvae. At this stage the pharyngeal glands of the workers produce "royal jelly" which is fed to very young larvae (for the first four days) and continually to larvae destined to become queens. After about two weeks, when the wax glands have developed, the workers' duties change to cell construction and cleaning, and receiving stores of pollen and nectar from foraging bees. Three weeks later the worker becomes a field bee with large pollen baskets on the hind legs. Early work-

ers live for about eight weeks but those emerging in late summer or autumn live over winter. Drones are driven out of the hive by workers in the autumn to conserve food stocks. The queen maintains cohesion in the colony by the production of chemicals called pheromones, which influence the behavior of the bees; these induce the workers to nurse the larvae. When this substance diminishes, as the queen ages, the workers start producing new queens by feeding certain larvae solely on royal jelly. At some stages in the colony's life the old queen leaves with about half the workers. At this time a new queen emerges and the other developing queens are usually killed. The new queen mates during a nuptial flight with the drones and returns to head the colony.

One of the most astounding features of bees is their ability to communicate the position of a rich food source to other worker-foragers, which is done by a special dance. On returning to the hive, a foraging bee alights on a vertical surface of the

Termite colonies are controlled by pheromones which are secreted by the queen. She lies in the colony, trapped by the size of her white abdomen — 5inches (12centimeters) long — from which she releases several thousand eggs a day. The eggs are fertilized by the only sexually active male — visible in the foreground — which is larger than the other termites. Except for her mate, the queen produces all the members of her colony which are blind, wingless and sterile. But at certain times of the year, she gives rise to winged, fertile offspring, which leave the colony.

comb and begins to dance in relation to the position of the sun. She moves in a straight line, waggling her body. If she moves upward the food source lies towards the sun; downward means that it lies away from it. Moving at angles to the left or right she indicates that the bees must fly with the sun on their left or right respectively. The speed of the dance indicates the distance of the food. The other workers follow the progress of the dance by touching her, and then fly off toward the food.

Ant colonies are similar to those of bees; there is usually one queen and many workers, with only a few males. But many ant workers are specialized to perform certain tasks only. Soldier ants, for example, have huge heads and enlarged mouthparts, and their function is to defend the colony. Other ants are workers, tending the larvae and collecting food. Unlike the bees, ant larvae seem to be able to develop into any caste (within the limits of sex), but development hinges on feeding — larger ant larvae develop into queens, for example.

The types of ant colony vary from species to species; they may be groups of hunters, food gatherers, farmers or even stockbreeders. Army ants are hunters, and swarm over victims in enormous numbers, killing them for food. Food gatherers collect plant material and bring it into the nest. Often these ants "milk" insects like aphids for the sugary fluid (honeydew) they secrete. Many ants are fungus growers; inside their colonies they provide suitable conditions for the growth of particular species of fungi upon which they feed. Leaf-cutting ants, for example, are fungus growers; the leaf material that they bring back to the nest provides the substrate on which the fungus spores are sown. Other ants keep aphids, which they bring into the nest in winter, and put out again in the spring onto suitable plants.

Some species of ants (for example, *Formica sanguinea*) take ants of other species, which become slave-workers for the colony. Control of the slaves and the colony is also generally achieved by pheromones. Ants have also been known to mutiny and leave their home nest to raid another, which they move into, keeping its original inhabitants captive as slave-workers.

Economic importance

Many insects are of benefit to humans. The silk moth *(Bombyx mori)*, for example, provides silk. It is obtained from the caterpillars of this moth, which are farmed in huge numbers and fed on mulberry leaves. At pupation they begin to exude fluid silk and spin a silken cocoon. To extract the thread, the caterpillar must be killed, and the cocoon unwound — each one yields about half a mile of fine silk fibers.

Another beneficial insect is the honey bee, which is often domesticated and kept in specially constructed hives from which honey and beeswax are obtained.

Fruit trees, shrubs and flowering plants depend upon insects for pollination and great care

has to be taken when spraying such plants to ensure that insects are not killed by the chemicals. Many other insects are useful because of their function as biological pest controllers. Many ladybird beetles (family Coccinellidae), for example, feed on aphids, which can severely damage both ornamental and food plants, particularly citrus fruits.

Unfortunately, there are some insects that are pests; swarms of grasshoppers can lay bare vast fields of crops, and the cotton-boll weevil *(Anthonomus grandis)* attacks cotton plants. But some insects are also dangerous because they are the vectors of disease. Malaria, for example, is transmitted by *Anopheles* mosquitoes. Other diseases transmitted by insects include yellow fever, elephantiasis, sleeping sickness and typhus. In attempting to control these insects and the diseases they cause the cost of drugs, vaccines and eradication programs is enormous.

Ants milk aphids for their honeydew which the aphids collect from plants and release through the anus. But this relationship is not one-sided — the ants build shelters for the groups of aphids and protect their larvae. The ants drink the honeydew in large amounts; they can do this because their cuticle can expand. The ants return to the nest and regurgitate the sweet fluid to the ant larvae or other ants.

Molting — called ecdysis — is an essential process for growing insects because their exoskeleton is rigid and does not permit much growth. The intervals between ecdyses are called instars. The praying mantis *(Mantis religiosa)* has from 3 to 12 instars and takes about a year for complete metamorphosis. The pressure that ruptures the cuticle is achieved by the insect swallowing air or water, and contracting muscles. It then wriggles out of the old skin (visible at the bottom of the photograph).

Vertebrates

Despite its appearance, amphioxus (above) is not a fish; it is a lancelet, a chordate animal whose shape reflects the basic vertebrate structure. It has a notochord and segmented muscles.

Animals with backbones, called vertebrates, form only a tiny fraction of all known creatures. In many environments, however, they exert an influence out of all proportion to their numbers, because they are in general larger than invertebrates and in many instances more mobile. Both size and mobility are made possible by a strong internal bony scaffolding, jointed to give flexibility and surrounded by muscles that hold it together and make it move.

Structure and development

As vertebrates ourselves, we tend to think that the possession of bones is the great divide between animals. This is not really so and vertebrates are in fact only part of a larger group, the Chordates (Phylum Chordata). Chordata animals have a stiff jelly-like rod, called a notochord, running longitudinally through the dorsal part of the body, and acting as an internal support to a series of muscles or muscle segments (myotomes). Above the notochord is a hollow nerve tube, which is usually folded at the anterior end to form the brain. Below the notochord lies the digestive tract. At some time in their lives all chordates have paired gill slits and a tail.

In true vertebrates the notochord is replaced by cartilage or bone. The latter is a strong, hard substance, formed mainly of calcium phosphate plus collagen and other protein fibers. Bone has great strength for its weight and is well suited to act as the internal support for the body. It is possible that it originated as a waste product deposited in the skin of prehistoric fish-like marine creatures. When these animals began to live in brackish and fresh water they faced a change in the balance of their mineral environment, and this favored the production of bone, because the materials of which it was composed could be recalled if necessary for metabolic use. This role as a reservoir of certain minerals is still an important, although little-considered, function of bone. The early freshwater fish-like vertebrates developed abundant external bone — an enormous store of mineral wealth — which served the additional purpose of protective armor. Such a bony shell is obvious in some contemporary vertebrates, such as tortoises and armadillos. Though less evident, it also exists in all land vertebrates in which the large and delicate brain is protected by the bony box of the skull.

The only direct descendants of the original bone-clad vertebrates still surviving are the lampreys (Petromyzonidae) and the hagfishes (Myxinidae), although strangely these have lost all trace of hard skeletal structures. They have a persistent notochord and a gristle-like skull. True fishes, however, have well-developed skeletons and, in most cases, armor made of fine slips of bone (scales) set in the skin. The sharks and their relatives have retained bone-based external armor, but also have an internal skeleton formed of cartilage. In other vertebrates skeletal cartilage is mainly a juvenile tissue, replaced by hard bone as the animals grow.

Vertebrate types

There are seven classes of vertebrates: Agnatha, such as lampreys and their relatives; cartilaginous fish, such as sharks; bony fish; amphibians; reptiles; birds; and mammals. The structure of fishes restricts them to water. During the late Devonian Period (about 400 million years ago), however, some fishes, stranded in drought conditions, struggled towards new pools using large, strong stiltlike fins.

Since that time vertebrate animals have become less and less dependent upon a water environment, though all still need water to live (even if they only drink it). First among these terrestrial creatures are amphibians, which must return to the ancient habitat to lay their eggs. Their young generally look like tiny fishes but the adults are free to colonize the land.

Vertebrates consist of seven classes: Agnatha, such as lampreys and their relatives; cartilaginous fish, such as sharks; bony fish, such as carp and bass; amphibians, such as frogs (above right); reptiles, such as the green turtle (left); birds, such as the sacred kingfisher (above right); and mammals, such as the oryx (right).

Next in evolutionary sequence came the reptiles. These developed shelled eggs, enclosing their embryos in their own private pool (the watery white of the egg) during the larval ("tadpole") stage. The egg also supplies the embryo's nutrition, in the form of the yolk, which is the perfect, complete food needed for growth to hatching point. Hatched reptiles show no trace of gills and their skin is hard and waterproof. Their metabolic pattern requires external warmth, however, so they are restricted to climatically favorable parts of the world.

In the Triassic Period, about 200 million years ago, mammals evolved from reptile species that are now extinct. Activity is the keynote of their being and mammalian bodies are generally well adapted for easy movement. Such activity must be fueled by abundant food, and the relatively high-energy metabolism of mammals generally requires that the body is maintained at a steady temperature (aided by insulating hair or fur) while regular breathing provides the large amounts of oxygen such metabolism needs. The young of almost all species of mammal are born at a relatively advanced stage of development, because a special maternal organ, the placenta, provides nourishment for the fetus while it grows within its mother's body. The placenta also removes excretory products. After birth, young mammals are fed on maternal milk which gives them all the nourishment they need at this stage. In many species, important maternal and social ties are formed and cemented during this period of suckling. Most mammals have relatively large and complex brains which enable the young to learn behavior, particularly from their parents.

During the Jurassic period, about 160 million years ago, birds — the last of the vertebrate classes — evolved from reptile stock far removed from the ancestors of mammals. They have exploited the possibilities of flight more fully than other vertebrates and to achieve this have developed the most energy-intensive metabolic system of all. Birds show many modifications of the general vertebrate pattern, the most obvious being the transformation of the forelimbs into wings. The body is maintained at a high temperature and is insulated by feathers. Reproduction still depends on reptile-like eggs which are laid individually and at intervals, because of the need to keep weight to a minimum for flight. Many birds care for their young for a prolonged period, which establishes social bonds comparable to those seen in mammals.

Fishes

Almost three-quarters of the Earth's surface is covered by water, in which live 21,000 known species of fishes. From their earliest jawless ancestors in the Silurian period, 500 million years ago, fishes have diversified to inhabit widely differing aquatic habitats. Sleek, fast-swimming tuna *(Thunnus* sp.) live in the surface waters of the oceans, while the dark abyssal regions are inhabited by lantern fish and other deep-sea forms. Freshwater contains fishes such as trout — able to survive in the violence of mountain torrents — and, at the other extreme, lungfish which inhabit temporary pools and can withstand several years of drought. Some fishes, like the blind cave characin *(Stygichthys typhlops),* have even more specialized habitats.

Despite their similar basic design, fish forms show considerable variety. One of the smallest fishes, the less than ½ inch (11 millimeters) goby from the Philippines *(Mistichthyes luzonensis),* is also among the smallest vertebrates. The basking shark *(Cetorhinus maximus)* grows up to 40 feet (12 meters) long and weighs up to 4 tons, and the whale shark *(Rhincodon typus)* can reach more than 60 feet (18 meters). Coloration also varies, from the dazzling multicolored fishes of coral reefs to the colorless cave fishes and the almost totally transparent glass catfish.

Fishes are vertebrates, with a backbone of either bone or cartilage. Among their distinguishing characteristics is the fact that they have fins, mostly in paired sets, and that their bodies are covered in scales. Most fishes breathe using internal gills, to which blood is pumped by a two-chambered heart (mammalian hearts, by contrast, have four chambers). They are poikilothermal ("cold-blooded"), because their body temperature varies according to the temperature of the water. Tropical fishes, therefore, do not in fact have cold blood at all. Most fishes are oviparous (egg-laying), though in some species the eggs are retained in the female until after they hatch (ovoviviparous), and in a few species, such as the Atlantic manta ray *(Manta birostris),* the developing embryos are nourished by internal secretions from the female before being born alive (viviparous).

Fishes are divided into two groups: bony fishes (class Osteichthyes) and cartilaginous fishes (class Chondrichthyes), according to whether the main skeletal material is comprised of bone or cartilage.

Most fishes are carnivorous, preying on smaller fishes or other sea creatures, as this grouper feeds on a shoal of dwarf herrings.

A manta ray *(Manta birostris)* feeds on plankton, which it funnels into its mouth with the two "horns" that are extensions of its snout. Its huge pectoral fins propel it through the water by flapping like wings. This photograph also shows several remoras *(Echeneis naucrates)* attached by their suckers to the manta's large gill openings, where they feed on plankton that the manta leaves behind and on parasites from the manta's skin.

The streamlined shape and the segmented body muscles of fishes allows them to use sinuous side-to-side movements of the body and tail to move through water.

Adaptations to aquatic life: swimming

Because of its higher density and viscosity, water offers more resistance to movement than air. To minimize drag, therefore, the basic fish shape is streamlined. Backward-pointing scales cover the body and the skin secretes slippery mucus to cut down water resistance further.

In most fish, propulsion comes mainly from the caudal (tail) fin. Muscle blocks on either side of the backbone contract alternately, thus causing the body to bend and the tail to flex from side to side. The tail fin forces the water backward and this propels the fish forward. Other fins control the direction of movement and help the fish stop. Pectoral fins act as elevators, adjusting the vertical pitch of the fish as it moves. Dorsal and anal fins, which are unpaired, prevent the fish from rolling around its long axis. The paired pelvic fins often act like rudders in changing direction.

In some species the tail has acquired other functions apart from providing propulsion. The sea horse (*Hippocampus* sp.), for instance, uses its tail to cling to weeds; it swims by lateral undulations of its dorsal fin.

Adaptations to aquatic life: buoyancy

Cartilaginous fishes are denser than water and so tend to sink. This is described as negative buoyancy. To overcome this tendency they use their pectoral fins, the front of their head and their tail fin to produce lift as they swim. When stationary, however, they still tend to sink.

In contrast, most bony fishes can give themselves the same density as water (that is, they achieve neutral buoyancy) and are therefore able to maintain their position in the water even when not swimming. They do this using their swimbladder, a gas-filled, lung-like sac, which can be inflated or deflated to alter the buoyancy of the fish, allowing it to rise or descend in the water. Predacious fish such as the pike *(Esox lucius)* exhibit this ability well, remaining stationary as they wait for passing prey, using only small fin movements to maintain their position.

Adaptations to aquatic life: sense organs

Most fishes living in clear waters have good eyesight. The archer fish *(Toxotes jaculator)*, for example, is able to aim a jet of water at insects above the surface, and even allows for the bending of the light rays as they pass from air to water.

Nostrils in the snout allow for the sense of smell. Sharks are renowned for their ability to detect blood in the water from as far away as one-third of a mile (half a kilometer). Many bottom dwellers, such as the catfish *(Siluriformes)*, possess whisker-like barbels around the mouth, which give them the senses of taste and touch.

Two other sensory systems are found exclusively in aquatic animals, and both relate to water's conducting properties. These are the pressure sense provided by the lateral line system, and an electric sense possessed by some species. In a few species the latter is also associated with an ability to generate an electrical pulse.

The lateral line system can detect pressure changes caused by disturbances in the water around the fish. In experiments, blinded fishes have been trained to locate a moving glass rod by this method. In the wild it enables a fish to detect moving food items, approaching predators, or other members of a shoal.

Many fishes that live in muddy water or are active at night also have an electric sense. Electric receptors can be identified in sharks, dogfish, catfish and mormyrids (Mormyridiformes). Of those species which also possess electric organs capable of generating electrical pulses, some, such as the gymnarchid *(Gymnarchus* sp.), can detect irregularities in the electric field and thus locate objects in the cloudiest water. The electric eel *(Electrophorus electricus)* has the well-known ability to produce much more powerful pulses (up to 500 volts) when alarmed, or to stun prey.

The basic vertebrate shape of a fish is apparent even in a flatfish, where the body appears to have turned onto its side. This computer-colored X ray of a plaice shows the fish's head, skeleton, internal organs and muscles.

A fish relies on its fins for movement, balance and position. The dorsal fins, on the fish's back, help to keep it upright and traveling straight, though some lateral "weaving" inevitably occurs. The pectoral fins, at the sides, control its horizontal position.

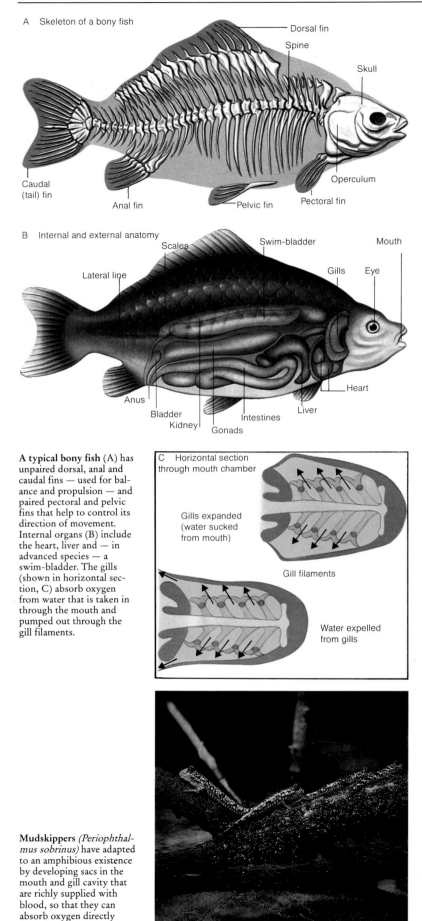

A Skeleton of a bony fish

Dorsal fin

Spine

Skull

Caudal (tail) fin

Anal fin

Pelvic fin

Pectoral fin

Operculum

B Internal and external anatomy

Scales

Swim-bladder

Mouth

Lateral line

Gills

Eye

Anus

Bladder

Kidney

Gonads

Intestines

Liver

Heart

A typical bony fish (A) has unpaired dorsal, anal and caudal fins — used for balance and propulsion — and paired pectoral and pelvic fins that help to control its direction of movement. Internal organs (B) include the heart, liver and — in advanced species — a swim-bladder. The gills (shown in horizontal section, C) absorb oxygen from water that is taken in through the mouth and pumped out through the gill filaments.

C Horizontal section through mouth chamber

Gills expanded (water sucked from mouth)

Gill filaments

Water expelled from gills

Mudskippers (*Periophthalmus sobrinus*) have adapted to an amphibious existence by developing sacs in the mouth and gill cavity that are richly supplied with blood, so that they can absorb oxygen directly from the air.

Adaptations to aquatic life: gills

The internal gills of fishes have a large surface area over which gaseous exchange between the water and the blood can occur. The gill area is related to the life style of the fish. Active species need more oxygen and so have large gills. Sluggish forms, like the bottom-dwelling toadfish (*Opsanus* sp.), have smaller gill areas in proportion to their body size. The mouth and gill (opercular) cavities form a mechanism that pumps water over the gills, a one-way flow being ensured by flaps of skin in the mouth. The gills are so arranged that the blood in the gill filaments and the water-carrying oxygen flow in opposite directions, and this counter-current system allows for the most efficient exchange of gases. In some sharks the respiratory current from the gill slits is strong enough to propel the shark forward.

Certain cells in the gills also help fishes to maintain their internal water and salt balance. In freshwater fishes these cells absorb salts from the water, to keep up the body's salt level, while in marine bony fishes other cells excrete salt to compensate for the high levels of salt swallowed in sea water.

Cartilaginous fishes

Because they lack bony skeletons yet possess gill slits, cartilaginous fishes (class Chondrichthyes) were originally considered a primitive group. They do not appear before bony fish in the fossil record, however, and zoologists now believe that bone has actually been replaced by cartilage in these fishes.

About 600 species of cartilaginous fish have been identified. Most are marine and carnivorous, although the largest representatives are actually harmless filter feeders. They show the complete range of reproductive strategies. Egg-layers lay large yolky eggs: the familiar "mermaid's purse" is the horny egg case of the dogfish. Offshore sharks are often ovoviviparous, the newly-hatched sharks feeding on unfertilized eggs within the female before birth. Some sharks, however, and some rays are viviparous, but in all cases fertilization is internal. The male fish has a pair of claspers formed from the pelvic fins, which are inserted into the female to transfer the sperm.

Cartilaginous fish are divided into two groups: the chimaeras (subclass Holocephali), and the elasmobranchs (subclass Elasmobranchii).

Chimaeras

There are about 30 species of these unusual deep-sea fishes. Chimaeras differ from other cartilaginous fishes in possessing only one pair of gill openings (elasmobranchs have at least five pairs). The upper jaw carries tooth plates and is fixed to the brain case, unlike the mobile jaws of sharks. Chimaeras have long thin tails and swim by flapping their large pectoral fins.

Elasmobranchs

Sharks, dogfish, skates and rays are the typical elasmobranchs. All have gill slits and, usually in front of these, another opening called the spiracle. The mouth, on the underside of the head, is filled with rows of teeth, which are developed from the toothlike structures that cover the body. In sharks only one row of teeth is used at a time, but as soon as teeth are lost or worn away, others move forward to replace them.

Skates and rays (order Rajiformes) have flattened bodies with the gill slits opening on the underside. Some sharks also have flattened bodies, but the gill slits are always on the side of the head. Many skates and rays are bottom-dwellers, often lying buried in sand with only the eyes protruding from the top of the head. As with chimaeras, the long thin tail is not used for swimming in most species. Instead they move by flapping or forming ripples in the very large pectoral fins. One of the largest rays is the Atlantic manta, measuring up to 23 feet (7 meters) from the tip of one pectoral fin to the other. Like the largest sharks it is a filter feeder, collecting small fish and crustaceans. Electric rays (suborder Torpedinoidea) use their electric organs (capable of delivering a charge of 200 volts) to stun their prey, or as a means of defense. The stingrays, sluggish bottom-dwellers, have a different defensive strategy — a poisonous spine at the base of the tail.

Most sharks and dogfish (order Selachii) have a streamlined, torpedo shape. Probably the most feared and the most likely to be dangerous to man is the great white shark *(Carcharodon carcharias)*. Specimens up to 21.5 feet (6.5 meters) have been reported. Stomach content analyses suggest their normal diet is fish, dolphins, sea lions and seals; but they are also known to attack bathers.

The smallest members of this group are the dogfishes (family Scyliorhinidae), many being smaller than 3 feet (1 meter) in length. Dogfishes live in shallow coastal waters, where they feed on mollusks, worms and other invertebrates.

Bony fishes
About 95 per cent of living fishes belong to the class Osteichthyes (bony fishes), making this the largest vertebrate class. As their name suggests, at least part of the skeleton is formed from bone, although cartilage often also occurs. Bony fishes

can be readily distinguished from cartilaginous fish by the presence of a bony gill cover (operculum). Also, the mouth is usually at the very front of the head, instead of on the underside as in cartilaginous fishes.

Reproductive strategies also differ from those of cartilaginous fishes. In contrast to the large yolky eggs of the latter, bony fishes tend to lay numerous small eggs — the number (which can be millions) usually relating to the hazards of the life style. Because the males do not possess claspers to transfer sperm, fertilization takes place externally.

Another typical feature is the air sac, which functions as a lung in the more primitive forms, and as a swim-bladder in the rest. Bony fishes are

Sharks are voracious carnivores, equipped with a large mouth and rows of sharp teeth to tear flesh from their prey. This photograph also shows the shark's gill arches in the sides of the mouth cavity.

Electric eel (from side)

Electric organ

Electric eel (from above) showing its electric field

Electroplates (magnified) in sequence

Electric fishes can not only sense electric currents in the water but can also generate an electric field and (in some species) strong electric pulses. The organs responsible consist of rows of electroplates that work neurochemically, intensifying what is essentially a nervous impulse. In the electric eel *(Electrophorus electricus)* these organs are located on either side of the body, and produce an electric field that resembles the magnetic field of a bar magnet.

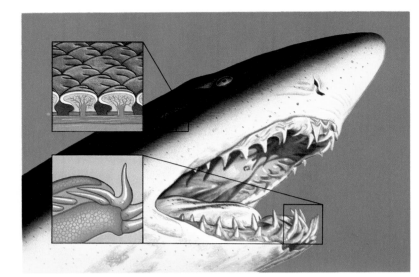

A shark's teeth grow in rows (lower inset). The illustration shows the teeth of a sand tiger shark *(Carcharias taurus),* which are typical of many shark species. A shark's placoid scales (upper inset) have a similar structure to its teeth, and this is apparent in the cross-section that shows the hard tooth-like covering and the soft core of pulp.

tors of tetrapods. Consequently most zoologists believe that land vertebrates arose from a freshwater group of lobefins. Today, lobefins are represented by a single species, the coelacanth *(Latimeria chalumnae),* although more than 50 fossil species are known. This "living fossil" has altered very little since it first appeared in the Devonian period, 400 million years ago.

The lungfishes are also a relict group. They were the most numerous fishes in the Devonian period, but there are now only six living species, distributed in Africa, Australia and South America. All have functional lungs, which allow them to survive in poorly oxygenated waters which occasionally dry up.

Ray-finned fishes

Most freshwater fishes and commercially exploited marine fishes are ray-finned (subclass Actinopterygii). As their name suggests they have fins supported by bony rays. They have two pairs of nostrils, but these do not penetrate to the mouth. Ray-finned fishes fall into four subdivisions: Polypteri, Chondrostei, Holostei and Teleostei. The first three are considered primitive forms.

Birchirs or reedfishes (infraclass Polypteri) occur in African inland waters. They possess a pair of lungs, so like the lungfishes they can survive in oxygen-deficient swamps. Sturgeons (infraclass Chondrostei) include the world's largest freshwater fish, the beluga *(Huso huso),* which can grow to a length of 14 feet (4 meters) and a weight of 2.850 pounds (1.300 kilograms). Sturgeons are also the sought-after source of caviar (which is salted sturgeon roe). Their skeletons are made mainly from cartilage, and they have several apparently primitive fea-

usually subdivided into three groups: lungfishes (subclass Dipnoi), lobe-finned fishes (subclass Crossopterygii) and ray-finned fishes (subclass Actinopterygii).

Lungfishes and lobefins

Zoologists have studied lungfishes and lobefins with particular interest since it was realized that similar species probably gave rise to the first land vertebrates. Lungfishes possess choanae (nostrils that connect the mouth cavity to the outside air) as in modern air-breathing amphibians. They and lobefins have paired fins containing bony supports and muscles in their bases, from which the weight-bearing limbs of land vertebrates could have developed. Since modern lungfishes have highly specialized jaws with tooth plates, however, they are unlikely to have been the ances-

Camouflage is the only means of defense for some fishes. The mottled skin and the flat shape of a plaice *(Pleuronectes platessa)* enable it to merge with the rocks and gravel of the sea bed. Plaice may also try to bury themselves beneath a sandy surface for even greater concealment.

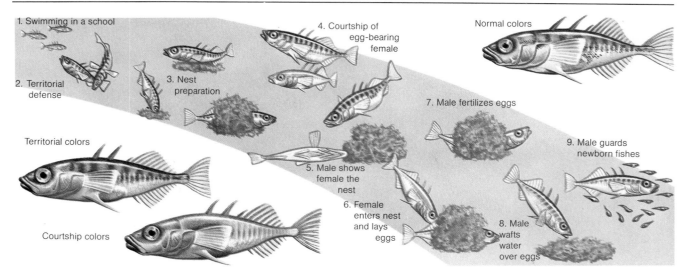

1. Swimming in a school

2. Territorial defense

3. Nest preparation

4. Courtship of egg-bearing female

Normal colors

Territorial colors

Courtship colors

5. Male shows female the nest

6. Female enters nest and lays eggs

7. Male fertilizes eggs

8. Male wafts water over eggs

9. Male guards newborn fishes

tures such as spiracles and an asymmetrical tail fin. They do not have a swim-bladder, however.

The North American gar (*Lepidosteus* sp.) and bowfin (*Amia calva*, infraclass Holostei) do have a lung-like swim-bladder and can breathe air when necessary. With sharp teeth, both are voracious carnivores preying on other fishes and often causing damage to commercial nets in the process.

Teleost fishes

By far the most numerous and diverse group of fishes are the teleosts (infraclass Teleostei). Perhaps more than to anything else, their success can be attributed to their mobility, achieved by the development of the swim-bladder, highly mobile fins and reduced scales.

The swim-bladder allows for buoyancy control, and within the group various stages of swim-bladder development can be seen. In the carp (*Cyprinus* sp.) the swim-bladder is still connected to the gut by a duct, and is filled by swallowing air. In the perch (*Perca* sp.) and other advanced teleosts, however, this duct has been lost, and the bladder is filled by gas drawn from the blood at a special "gas gland."

The fins of teleosts also enhance their mobility. The tail fin is symmetrical, giving effective forward propulsion. Another feature that can be observed in this group is the tendency of the paired pelvic fins to move forward to underlie the pectoral fins in more "advanced" species. In the salmon (*Salmo salar*) the paired fins are well separated, whereas in the perch the pelvic fins have moved right forward. Again this appears to aid mobility, since stopping and turning are easier with this arrangement.

Other teleost features include a reduction of the bone in the scales to a very thin layer. Two types of teleost scale occur: cycloid scales as in the carp, and ctenoid scales as in the perch. Both types are considerably lighter than the thick ganoid scales of primitive bony fish. Scales are of considerable interest to fisheries biologists in temperate regions, because the age of the fish can be determined by counting the growth rings on its scales.

Teleosts are divided into about 30 different orders, of which the most important are eels; salmon, trout and pike; carps, characins and catfishes; codfishes, spiny-finned fishes; and flatfishes.

Eels

Characterized by their snake-like bodies and smooth, slimy, often scaleless skin, eels (order Anguilliformes) are easily recognized. They have no pelvic fins, and their anal and dorsal fins have fused to form a fin seam, which is used in swimming.

Eels have an unusual transparent larval form called a leptocephalus (leaf-head), which bears little resemblance to the adult, and was originally considered a completely different fish. In the North Atlantic eel (*Anguilla anguilla*) the whole

In stickleback courtship the male acquires a red belly as he establishes his territory and prepares the nest. The colors intensify as he identifies a female by her egg-swollen shape, then encourages her to lay her eggs in the nest. Once fertilized the eggs are tended by the male, which continues to protect the young when they have hatched.

The black marlin (*Makaira indica*) is the largest of the billfish, sometimes exceeding 13 feet (4 meters) and 1,500 pounds (700 kilograms).

reproductive cycle is extraordinary. Adult eels change from freshwater to saltwater fishes, then migrate from Europe to spawning grounds in the Sargasso Sea, more than 3,350 miles (5,400 kilometers) away. Once hatched, the larvae make the return journey, aided by the Gulf Stream. Eventually — possibly after as much as three years — they enter European rivers, often at night, where they grow into freshwater, adult forms.

Most eels are marine, however, like the moray *(Muraena helena)* and conger eel *(Conger conger)*, both of which are voracious fish eaters. Garden eels *(Gorgasia maculata)*, in contrast, live in tubes and filter tiny zooplankton for food. Despite its name and shape, the electric eel is not a true eel, being more closely related to the carps.

Salmon, trout and pike

The order Salmoniformes is typified by salmon and trout, both migratory fishes from northern temperate areas, which often use offshore feeding grounds yet return to freshwater to spawn. Lesser known species include the char *(Salvelinus* sp.), grayling *(Thymallus* sp.), smelt *(Osmerus* sp.) and whitefish *(Coregonus* sp.). All can be recognized by their large mouths and the presence of a small, rayless adipose fin between the dorsal and tail fins. All are popular food-fishes.

The pike *(Esox* sp.) is also a member of this order. A large mouth filled with sharp teeth enables a pike to feed on other fish and even small mammals and birds taken at the water's edge or surface.

Carps, characins and catfish

The majority of the world's freshwater fishes, more than 3,500 species, belong to the same group as carps and catfishes (order Cypriniformes). All possess a swim-bladder connected to the gut, cycloid scales, and well-separated pectoral and pelvic fins.

The 2,000 species of carp-like fishes include most European coarse fishes, such as carp *(Cyprinus* sp.), roach *(Rutilus* sp.), bream *(Abramis* sp.),

dace *(Leuciscus* sp.) and tench *(inca tinca)*, plus the non-European goldfish *(Carassius* sp.). Carp lack jaw teeth, but possess teeth on the pharyngeal (throat) bones. In the common carp *(Cyprinus carpio)* these crush the plant material and invertebrates on which it feeds. The mouth can also be protruded, which allows carp to suck up food particles effectively. Originally from central Asia, the common carp has been widely distributed and farmed by man. Recently another species, the herbivorous grass carp, has been introduced into several areas to help control weeds, which it eats.

Unlike the carps, characins, which include the piranha *(Serrasalmus* sp.) and tiger fish *(Hydrocinus* sp.), have well-developed teeth. These are brightly colored fishes from South America and Africa respectively. Many species of characins, however, like the tetra *(Hemigrammus* sp. or *Hyphessobrycon* sp.) have been distributed worldwide to tropical fish enthusiasts.

Catfishes (order Siluriformes), in contrast, are not colorful, being mainly nocturnal bottom-dwelling fishes. Consequently their eyes are small, but they bear up to four pairs of conspicuous barbels around the mouth, and these are sensitive to touch and taste.

All of this group of carps, characins and catfishes (superorder Ostariophysi) share two interesting features. The first is that they possess a series of three bones (the Weberian ossicles) connecting the swim-bladder to the inner ear. Sound vibrations in the water are picked up and amplified by the gas-filled bladder and then transmitted to the ear, giving these fish excellent hearing. To match this, some catfish can also produce sounds, by drumming on the swim-bladder or scraping the spines of the pectoral fins. The second is that they all have a similar "fright" reaction. When one fish is injured its skin cells release a chemical into the water. Other individuals react to this by either fleeing or hiding, so that if a predator attacks one member of a shoal, the others escape.

Codfishes

Another commercially important group, including the Atlantic cod *(Gadus morrhus)*, haddock

Moray eels *(Gymnothorax* sp.) are found in subtropical and tropical seas, where they hide among rocks and in crevices except when hunting for food. Most morays are aggressive; all have sharp teeth; and some species have poison glands in the mouth which pour poison into the wound caused by the moray's bite and so disable or kill the prey.

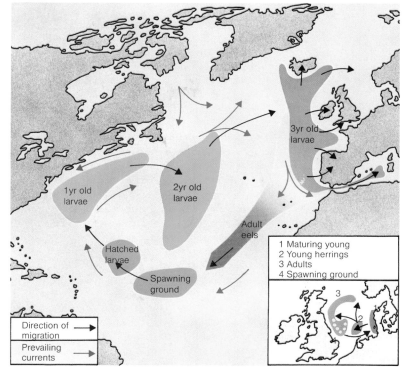

Horse mackerel (Trachurus trachurus, above) have streamlined bodies and powerful tails. They swim in schools in search of the smaller species on which they prey, and are an important food fish.

Eels (Anguilla anguilla) migrate vast distances to spawn (right), and their larvae return slowly to the European rivers. Herrings migrate over short distances — the inset map shows North Sea shoals.

1yr old larvae

2yr old larvae

3yr old larvae

Adult eels

Hatched larvae

Spawning ground

1 Maturing young
2 Young herrings
3 Adults
4 Spawning ground

Direction of migration

Prevailing currents

(Melanogrammus aeglefinus) and hake (Merluccius merluccius), are known collectively as codfishes or gadids (order Gadiformes). They are essentially cool-sea fishes, living mainly on the continental shelves in the Northern Hemisphere. Hake, however, also live in deep water and have invaded subtropical areas. At night they surface to feed on other fish such as herring. Growing up to 1 foot (30 centimeters) in length, hake now form the main catch of fishing industries in Europe, South Africa and South America.

Spiny-finned fishes

The most advanced teleosts have spiny rays in the dorsal, pelvic and anal fins (as anyone who has handled a perch carelessly will know). Advanced features include ctenoid scales, a swim-bladder not connected to the gut, and pelvic and pectoral fins sited close together.

There are more than 7,000 species of perch-like fish (order Perciformes), which include the angelfish (Pomacanthus sp.), butterflyfish (Chaetodon sp.), mackerel (Scomber scombrus) and tuna (Thunnus thynnus). They are active, often brightly colored fishes, with good eyesight and color vision.

The European perch (Perca fluviatilis) is an active predator, feeding on other fish — as is the much larger Nile perch (Lates niloticus) — but within this group all manner of feeding specializations are found. One family, for example, the cichlids (Cichlidae), are particularly diverse. In Lake Malawi, 200 different species of cichlids coexist, feeding according to their separate tastes on mollusks, insects, fishes, eggs, plankton, algae or even the scales of other fishes. In each case, the pharyngeal teeth appear specially adapted to the diet.

Flatfishes

Many teleosts are laterally compressed, but this is taken to extremes in the flatfish (order Pleuronectiformes).

The larvae of flatfish are conventional in appearance, but as they grow they gradually become compressed and turn over on one side. The eye on the lower surface migrates to the upper surface, the underside becomes white, and the upper surface darkens.

To match this shape, most flatfish live on the sea bed in coastal waters, and some — like the plaice (Pleuronectes platessa) — possess the ability to alter their coloration to match their surroundings. Commonly-eaten flatfish such as sole (Solea solea) are small, although others, such as halibut (Hippoglossus hippoglossus), reach weights of 400 pounds (180 kilograms).

The coelacanth (Latimeria chalumnae), a "living fossil," may be a link between fishes and amphibians. Once thought to be extinct for 60 million years, it was discovered in 1938 in the waters off southern Africa.

Frogspawn consists of the fertilized eggs of a frog, with the dividing cells surrounded by a gelatinous envelope. The eggs of other egg-laying amphibians, such as toads, newts and salamanders look similar.

Amphibians

The first fossil evidence of a terrestrial vertebrate — an amphibian, *Ichthyostega* — is found in the Devonian Period, between 390 and 350 million years ago. The Devonian Period was a time of great ecological changes, which resulted from massive disruptions of the surface features of the earth. For the evolution of amphibians, the most significant of these were changes that altered the level of the seas.

As water levels rose then receded, the sea left behind organic materials in which plants could thrive. This tended to encourage the development of areas of lush vegetation in swamplike coastal regions. These conditions favored creatures that could obtain their oxygen from air as well as from water and it is probable that it was in such an environment that the first amphibians evolved.

In prehistoric times there were at least six orders of amphibians (class Amphibia) but now only three orders remain: newts and salamanders (Urodela); frogs and toads (Anura); and caecilians (Apoda). Although the majority of amphibians inhabit the tropics, representatives of the class are found throughout the world, except in places where there is no water at all or where there is permanent frost. In cold climates they hibernate during winter. Like most animals except birds and mammals they are cold-blooded (poikilothermal) and their body temperature changes with fluctuations in the temperature of their environment.

Amphibians vary in length from an inch or less (for instance African sedge or reed frogs of the *Hyperolius* genus) to more than 5 feet (1.5 meters), the largest amphibian being the Japanese giant salamander *(Megalobatrachus japonicus)*, although a caecilian from Colombia *(Caecilia thompsoni)* reaches nearly the same length. Amphibians have a wide flat skull attached to the spinal column. The latter may be short, as in frogs and toads which usually have eight vertebrae in the trunk, or extremely long, with as many as 250 vertebrae in some salamanders and caecilians. The urodeles and anurans have four limbs, with four fingers on each forelimb and five toes on each hindlimb. The apods have lost their limbs. Unlike the urodeles, adult anurans have no tails.

The skin of all amphibians is toughened (cornified) on the animal's upper surface and smooth on the lower surface. All shed their skin regularly — this molting is under hormonal control. The skin produces slimy or poisonous secretions which make these animals unpalatable to most predators and so afford protection. In many species the skin color can change, usually for camouflage or mating purposes. The skin also plays a part in respiration.

Respiration

In evolutionary terms one of the most profound modifications needed for animals to survive the emergence from an aquatic environment was the ability to absorb oxygen from air rather than from water. In amphibians, perhaps reflecting stages in this transition, respiration takes place in the gills, lungs, lining of the mouth or the external skin, alone or in combination depending on the species. The role of the skin in respiration varies from some species for which it is not particularly important, to others — for example some lungless salamanders (such as the European Alpine salamander, *Salamandra atra*) — in which the skin is the sole means of absorbing oxygen from the aerated water of the mountain streams in which they live.

Senses

Different amphibians have differing sensory needs. In almost all species the sense of touch is well developed but the development of organs for sensing things at distance — for sight, hearing and smell — varies greatly. Cave-dwellers such as the olm *(Proteus anguinus)* have little need of sight and their eyes are vestigial. Burrowing caecilians *(Ichthyophis glutinosus* from Sri Lanka, for example) also have only rudimentary eyes, although they have another organ like a small feeler which is associated with eye muscles and seems to have a sensory function. In other species, particularly frogs, the eyes are well developed.

A frog is called a tadpole in its larval stage. At the start of metamorphosis tadpoles have an ovoid body and head, and a tail. Back legs develop first, by about eight weeks, then front legs emerge from beneath the gill covering (operculum). By about 12 weeks the larva has lost its gills and gained the legs of the terrestrial adult form, but still has the tail of the aquatic form. Gradually this tail is reabsorbed as the frog assumes its adult shape.

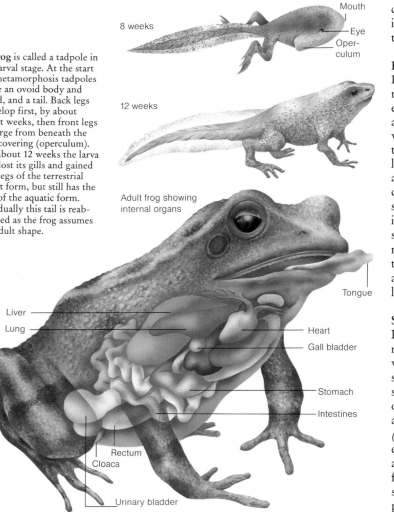

8 weeks

Mouth

Eye

Operculum

12 weeks

Adult frog showing internal organs

Liver

Lung

Tongue

Heart

Gall bladder

Stomach

Intestines

Rectum

Cloaca

Urinary bladder

Underground and cave-dwelling amphibians tend to have a good sense of smell, and this is also important for aquatic species, unless they live in clear water.

Lateral-line organs are present in aquatic species although they are less developed than in fishes. They appear as lines of little pits along the head and body, distributed in relation to the lateral-line nerves, and respond to vibrations or changes of pressure in the water, enabling the animal to control its equilibrium and posture and to detect predators or prey.

The existence of a sense of hearing in amphibians seems to depend on whether or not the animal has a voice. Urodeles and apods have neither voice nor visible ears, but appear to be able to detect vibrations. Frogs and toads, in contrast, have remarkably loud, even strident voices, and have an eardrum just behind each eye.

Movement

In water, amphibians can both walk and swim. During courtship, for example, newts often walk along the bottom of a pond. Frogs and toads use their hindlimbs for swimming, whereas newts propel themselves with their tails, keeping their hindlegs pressed together. Caecilians swim like eels. Although urodeles can walk on land, their tails and short legs restrict their mobility. Frogs and toads crawl on land, but also use their powerful hindlimbs for leaping — several species can jump 20 times their body length. The limbless caecilians usually burrow below ground; when observed above ground, however, their movements resemble those of a snake.

Feeding

Apart from anuran larvae (tadpoles), amphibians are carnivores, their main sources of food being insects and small invertebrates. Caecilians have a more varied diet, which includes other amphibians, fishes, and even some reptiles, small mammals and birds.

Amphibian teeth have no roots and grow all the time as they are worn down. Some urodeles have fixed, immobile tongues but others, such as the cave salamander *(Hydromantes genei)*, have tongues that can protrude to catch prey. This abil-

ity is particularly marked in frogs and toads, in which the tongue is attached at the front of the mouth. As the tongue flicks forward it scrapes the roof of the mouth, collecting a sticky covering which adheres to the prey, so that it can be drawn back to the anuran's mouth.

From the mouth food passes to the stomach where secretions from stomach glands (present in all amphibians though not in all fishes) start the digestive process. This continues through the intestine, aided by enzymes from the pancreas and bile from the gall bladder, to the rectum and cloaca. Urine also empties from the urinary bladder into the cloaca, from which waste products (urine and feces) are expelled.

Reproduction and life expectancy

Most amphibians lay eggs, and fertilization usually takes place outside the body. The young

Frogs have adapted successfully to many environments. The Costa Rican flying frog *(Agalychnis spurelli)* is a tree frog with webbed feet that enhance its ability to leap to such an extent that it appears to fly.

The fire salamander *(Salamandra salamandra)* protects itself with poison glands in its skin. These secrete an irritant fluid that burns the mouth of any predator that bites the salamander.

Amphibians are carnivorous and some are agile predators. In this photograph a palmate newt *(Triturus helveticus)* has captured a fish.

The voice of a frog is produced when air is pumped over the vocal cords in the throat. Male frogs of some species — in this photograph the painted reed frog *(Hyperolius viridiflavus)* — have a vocal sac which can be distended to produce a resonating chamber that intensifies the sound considerably.

pass through a larval stage, although in tailed amphibians and caecilians the distinction between this and the adult form is less obvious than it is in anurans. Some species (most caecilians, for instance) bear live young, others lay eggs that contain partially developed young, and others again (such as the common European frog, *Rana temporaria*) produce spawn in which the whole process of development occurs from the first cell division onward.

In captivity amphibians have been known to live for 15 years or more, but their life-expectancy in the wild is usually much less. Among the longest-lived are some slender salamanders*(Batrachoseps* sp.) and cave salamanders *(Hydromantes geneii)* which live for as long as 10 years in their natural habitat.

Frogs and toads
The anurans, loosely called frogs and toads, form the largest of the three orders of Amphibia, with some 2,700 species that have adapted to a wider range of habitats than the other orders and live on every continent except Antarctica. Although most are terrestrial rather than aquatic, some, such as the tree frogs (Hylidae), are primarily if not exclusively arboreal. In some of these species eggs are laid in water trapped among leaves, but some build nests, and some spawn in ponds and similar locations then return to their arboreal habitat.

The terms frog and toad are based on appearance and do not relate to actual phylogenetic distinctions. Anuran classification is based primarily on skeletal features, such as the presence of ribs, which distinguishes lower from higher anurans, or the structure of the pectoral and pelvic girdles. Beyond such relatively simple criteria anurans show enormous variety. There are anatomical variations in the extent to which hands and feet are webbed, in the shape of and function of fingers and toes, and in the distribution or arrangement of teeth, some species having none, others having teeth in the upper jaw only, and one genus *(Amphignathodon)* having teeth in both jaws.

Most anurans are nocturnal, although newly metamorphosed frogs take several days to assume their nocturnal habits. They are mainly dependent on their eyes for catching prey — their visual system responds to small, irregularly moving objects, and they will not react to even a preferred food source unless it moves.

An important feature distinguishing anurans from other amphibians is their voice. Almost all anuran males can increase the volume of the sounds they make by using vocal sacs in the mouth. The mating call of the natterjack toad *(Bufo calamita),* for example, can be heard more than half a mile away. Females, which are generally larger than males, have a quieter call.

In the breeding season the male anuran's call serves to attract females and enables a female to recognize a male of the same species. When a male anuran is ready to mate he will jump upon anything of the appropriate size, and will often clasp a male or some other inappropriate object (even a piece of mud or earth) before managing to find a female.

In most frogs and toads eggs and seminal fluid are emitted at the same time; fertilized eggs are then deposited singly or in clumps or strings. The majority of European and North American anurans lay clusters of hundreds of eggs, each in its own gelatinous envelope; the frogspawn of *Rana temporaria* is the most familiar example. One exception is the North American tailed frog *(Ascaphus truei),* which lays eggs that have been fertilized internally. Internal fertilization also occurs in one genus of toads *(Nectophrynoides)* found in Africa, which do not lay eggs at all but give birth to live young.

Salamanders and newts
The tailed amphibians (Urodela) make up the second largest of the three orders of Amphibia. There are about 330 species of salamanders and

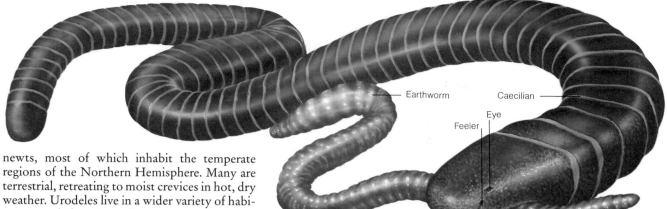

Earthworm — Caecilian —

Eye

Feeler

Mouth —

newts, most of which inhabit the temperate regions of the Northern Hemisphere. Many are terrestrial, retreating to moist crevices in hot, dry weather. Urodeles live in a wider variety of habitats nevertheless. Some live in trees, some never leave the water, and some live in the total darkness of deep caves. Like frogs and toads, the tailed amphibians need moisture; insects, worms or even fishes for food; and (for some species) fresh water in which to lay eggs.

Mating behavior differs from that of anurans. Salamanders and newts have no voice, so mating display is all-important. A male salamander's particular pattern of courtship behavior and coloring elicits a response only from a female of the same species. Male alpine newts *(Triturus alpestris)*, for instance, develop crests and striking coloration — blue on the back and flanks, orange on the belly — in the breeding season. Many species indulge in complex courtship rituals, ranging from lashing the tail to waft odorous secretions toward the female, to swimming back and forth in front of her.

During mating the male deposits a spermatophore (sperm packet), which the female takes up into her cloaca. The sperms are then released as the female releases her eggs. When these hatch the larvae resemble adults, although they have gills that disappear during metamorphosis.

Caecilians

The third amphibian order, and the one about which least is known, is the Apoda. Apods, or caecilians, have no limbs and look like large worms or smooth snakes. Lengths vary from just over 2 inches (6 centimeters) to nearly 5 feet (1.5 meters). Nearly all species live in warm-temperate and tropical regions. They usually burrow beneath the ground, seldom being seen above ground in daylight, although it is believed that most come to the surface at night. Eyes are of little use in such a habitat but caecilians have developed a sensitive "feeler" or tentacle which probably helps them search for the worms and insects that are the main constituents of their diet. Reproduction is by internal fertilization, and it is believed that caecilians either lay eggs or retain the eggs until the young hatch.

Caecilians are sometimes mistaken for very large earthworms, but when the two are compared — as in this illustration of a South American caecilian *(Siphonops annulatus)* eating an earthworm — there are obvious differences. Most significant of these are the caecilian's flattened head; its small, almost vestigial eyes, the small feeler located between its eye and nose, and its mouth. Some caecilians are strikingly colored: the species illustrated is blue-black, darker above than below. Others are mottled or striped.

In courtship many species of newts develop striking coloration. In some species the crest of the male also enlarges. Both characteristics are evident in this photograph of great crested newts *(Triturus cristatus)* mating.

Reptiles

Reptiles were the first true land-dwelling vertebrates, appearing more than 275 million years ago. They dominated the Earth from about 225 to 65 million years ago but today only four reptilian orders remain: the turtles and tortoises (order Chelonia); lizards and snakes (order Squamata); the crocodiles and alligators (order Crocodilia); and the tuatara (order Rhynchocephalia).

The most noticeable characteristic of reptiles is their outer covering of scales. They also lay tough, leathery eggs and are "cold-blooded" (that is, they lack an internal means of controlling body temperature). Because of this last feature most of the 6,000 or so reptilian species live in the tropics or subtropics, although some live in temperate regions, for example, the tuatara *(Sphenodon punctatus)* and the viviparous common lizard *(Lacerta vivipara)*, which is found as far north as the Arctic Circle.

The reptilian body

A reptile's body is covered by a dry, relatively thick waterproof skin, which helps to protect the internal structures and prevents dehydration. The scales on the reptilian body are really folds in the skin, composed of a hard, transparent dead material called keratin. In chelonians and crocodilians the outer keratin layer is continually worn away but is always underlain by a new layer. In snakes and lizards, however, the scales are shed at regular intervals. Lizards often shed their scales in strips, but some, like snakes, shed the whole skin in one piece, a process known as sloughing. The scales of many lizards and some crocodilians contain small plates of bone called scutes, which give the skin additional toughness.

In most vertebrates, including reptiles, the skeleton consists of bone and cartilage. The amount

The dinosaur Stegosaurus regulated its body temperature by means of the blood-rich skin that covered the diamond-shaped plates on its back. The blood absorbed heat from the Sun, carrying it around the body. In the shade the blood was cooled.

The reptilian skeleton varies greatly. The crocodilians have abdominal and thoracic ribs, and the chelonians have their ribs and vertebrae fused to their shell, but the ribs of the snakes extend to the end of the tail.

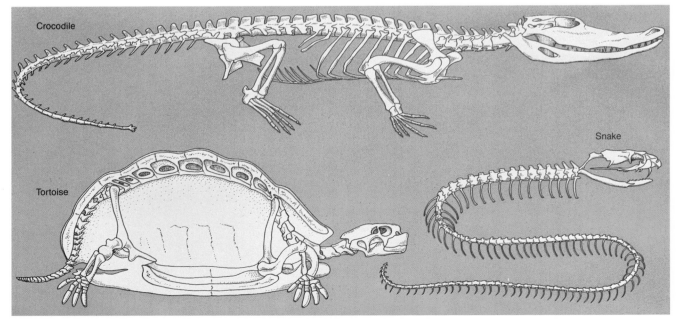

Crocodile

Tortoise

Snake

of cartilage in the body is highest in young reptiles, but as the reptile grows and becomes an adult, the cartilage is usually present only over the joints, the shoulder girdle and parts of the skull. Unlike mammalian bones, those of a reptile continue to grow throughout the animal's life.

The ribs of reptiles extend the length of the body, from the third vertebra of the spine to the beginning of the tail, whereas in mammals they are confined to the chest region. The skull consists of many separate bones. Similarly, the lower jaw is formed from several distinct but joined bones. The teeth are peglike or pointed and are all alike; they are not differentiated as are the teeth of mammals.

The reptilian brain is small in relation to the body — in many, it constitutes less than 0.05 per cent of the total body weight — and reptiles are relatively unintelligent, having very little ability to learn or adapt to new situations. But they compensate for this deficiency by possessing elaborate patterns of instinctive behavior that enable them to survive and reproduce successfully.

Reptiles breathe with the aid of two lungs, one or both of which may be well developed. Some species can also absorb oxygen in the water through the membranes that line the mouth and the genital and waste-eliminating chamber called the cloaca. The cloaca is at the end of the large intestine, and opens under the body in front of the tail. It receives eggs or sperm from the sex glands as well as feces from the intestine and urine from the kidneys. The water needed for elimination of waste products is reabsorbed into the blood through the walls of the large intestine and cloacal chamber. The anal waste is solid or semi-solid and there is little or no liquid urine. In this way, water is conserved in the body, a feature that is important for those reptiles that live in dry environments.

Senses and sense organs

Most reptiles have good eyesight. The fields of vision of each eye overlap to some extent, so that some of these animals have binocular vision, although in varying degrees. It is particularly highly developed in chameleons (*Chamaeleo* spp.), which must be able to judge distances accurately in order to catch their prey.

Relatively little is known about hearing in reptiles, although this faculty does not seem to be as important as vision. Like birds, mammals and amphibians, reptiles have a small bone (the stapes) which transmits sound vibrations from the eardrum to the inner ear. Most reptiles, apart from snakes and a few lizards, have external ears with eardrums on the surface or sunk down on a tube at the back of the head. However, they lack earflaps, except for crocodiles. Snakes also lack the middle ear cavity — they "hear" a sound mainly by picking up vibrations from the ground, which are then transferred from the lower jaw bone to the skull.

Smell is an important sense in most reptiles. Most possess special structures known as the

Reptilian eyes are varied. Chameleons' rotating eyes (A) allow them to keep still while watching for prey. Their eyelids cover most of the eye, which sharpens their focus. The round pupils of some lizards (B) contrast with those of the tokay gecko (C) which are vertical and constricted in four places to protect them from dazzle and improve vision. Crocodilian eyes (D) are covered with a membrane which allows them to be submerged in water. Tortoises (E) have round pupils protected by thick, folding eyelids.

Many species of lizard have spines, shields or frills on their body or head — the Bearded Dragon Lizard (*Amphibolurus barbatus*), for example. These features are either modified scales or skin membranes. They are puffed up to give the lizards an appearance of greater size in displays of aggression or in courtship.

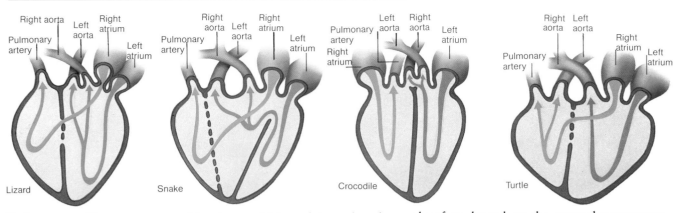

Lizard Snake Crocodile Turtle

The heart system differs among reptiles. The lizards have a perforated septum separating the ventricles. Non-aerated blood flows through it from the left ventricle into the pulmonary artery in the right. In the snakes the pulmonary artery, the aortas and the right atrium are in the right ventricle. A muscular ridge formed during contraction guides the non-aerated blood into the pulmonary artery. The septum of the crocodiles is whole and the two aortas come out of each of the ventricles. This is also true for chelonians but, because their ventricles are not perfectly divided, non-aerated and aerated blood flows into the aortas, so a mixing of blood occurs.

organs of Jacobson, which work in conjunction with the tongue as well as with the sense of smell. These organs are found above, and open into, the roof of the mouth and are partly lined with sensitive membranes similar to those that line the nose. As the tongue moves in and out it picks up minute particles from the air or off the ground, which are carried to the organs of Jacobson.

An additional sense organ — the pineal eye - occurs in the tuatara and many lizards (and was also present in many extinct reptiles and amphibians). It is found at the top of the head and opens, by means of a small hole, into the skull. It is covered by skin, which usually has a lighter color than the surrounding skin. The exact function of this "third eye" is unknown but it is light-sensitive and probably helps the animal to regulate its exposure to the Sun, and thus to control its internal body temperature.

Temperature regulation

The term cold-blooded is misleading when applied to reptiles because under certain circumstances a reptile's body temperature may be very high, even higher than the normal body temperature of humans 98.6°F (37°C). Reptiles are, in fact, poikilothermal — they do not have an internal mechanism to regulate body temperature — and

therefore depend on the external temperature. Body temperature is important because it is related to activity. Under extreme conditions of heat or cold the body becomes sluggish or even torpid because vital activities such as heartbeat and breathing slow down. But within a narrow temperature range — which varies from species to species — a reptile reaches its highest level of activity. For most the optimum temperature is between 77°F (25°C) and 99°F (31°C).

Reproduction

Most reptiles that live in temperate and subtropical climates breed in spring. As in most other animals, the urge to mate is triggered by an increase in the hormone levels in the sex glands and other internal organs. These changes are usually a response to changes in the environment, such as lengthening of the day, an increasing abundance of food, and a rise in air temperature. Tropical reptiles may breed several times a year, but even in these species mating usually follows a seasonal cycle.

With the exception of the tuatara, which mates by pressing its cloaca against that of the female, all male reptiles possess a penis through which sperm is introduced into the female's cloaca. Chelonians and crocodiles *(Crocodylus* spp.) have a single

Basking on warm rocks, reptiles such as marine iguanas *(Amblyrhynchus cristatus)* seek to prevent themselves from becoming too cold. They move into shade periodically to stop themselves from overheating. This dependence on external temperature is essential because they have no internal mechanism to regulate their body temperature. In addition, some reptiles change the color of their skin to make it more or less heat absorbent.

penis, whereas lizards and snakes have two, known as hemipenes, only one of which is used during copulation.

Most reptiles are oviparous — that is, the young hatch after the eggs have been laid. But some species of snakes and lizards, such as adders *(Vipera berus)* and the common lizard, are ovoviviparous (the young hatch while the eggs are still inside the mother's body). In such reptiles the young may still by surrounded by embryonic membranes when they are born. Exceptions to both these methods of reproduction are found in certain skinks (Scincomorpha) and snakes, in which the relationship between mother and embryo is more intimate, resembling that of mammals. In these reptiles a type of placenta develops which lies close to the lining of the mother's oviduct.

Tortoises, turtles and terrapins

Tortoises, turtles and terrapins (order Chelonia) are differentiated in the following way: tortoises live on land, turtles live in the sea, and terrapins live in fresh water. In the United States, however, most chelonians are commonly referred to as turtles.

The most striking feature of this order is the presence of a shell. The shell is composed of an outer layer of horny plates called laminae, and a thick inner layer of bony plates. Laminae arise in a layer of living tissue which lies between the laminae and the bony plates. The outer, horny layer increases in size with age, but its growth is arrested during hibernation. Each dormant phase produces a ring around the previous year's growth, similar to a growth ring in a tree. The laminae of the hawksbill turtle *(Eretmochelys imbricata)* are the source of the commercial material known as tortoiseshell.

Chelonians do not have teeth; instead they have a horny beak on the upper and lower jaws. Tortoises are mainly herbivorous, turtles eat a mixture of plant and animal food, and terrapins tend to be carnivorous, feeding on invertebrates and fish.

There are two suborders of chelonians, distinguished by the way the head is withdrawn into the shell. The Pleurodira are the side-necked turtles, found in South America, Africa and Australasia, and are a small group of two families. An example is the South American matamata terrapin *(Chelys fimbriata)*, which has a long snout with nostrils at the end and a long flap of skin behind each eye. It feeds on fish and other small aquatic animals.

The Cryptodira, or hidden-necks — a group that comprises seven families — bend the neck vertically up and down. The family Testudinidae in this group includes the land tortoises which are some of the most familiar chelonians, with about 40 species. The family Emydidae, with about 80 species, includes terrapins and pond tortoises. The pond tortoise *(Emys orbicularis)* and the red-eared terrapin *(Pseudomys scripta)* are commonly kept as pets.

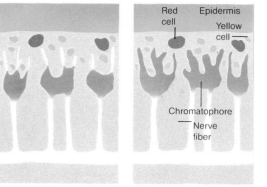

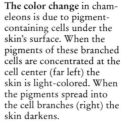

The color change in chameleons is due to pigment-containing cells under the skin's surface. When the pigments of these branched cells are concentrated at the cell center (far left) the skin is light-colored. When the pigments spread into the cell branches (right) the skin darkens.

The family Chelydridae is found only in North America and is comprised of the musk and mud turtles and the snappers. The common snapper *(Chelydra serpentina)* is a ferocious freshwater turtle that bites sharply when disturbed. Unusually, it cannot withdraw its head into its shell, and its tail, legs and neck are almost completely exposed on the underside. The musk turtles *(Sternotherus* spp.) are exceptional among reptiles in having scent glands on their body; they are often called "stinkpots" because of the musky odor they emit when handled.

Most reptiles lay their eggs on land, usually concealed under a stone or buried in soil or sand. But some, such as the rat snakes *(Elaphe* spp.), lay them in tree hollows. The eggs are protected by a relatively tough, parchment-like shell.

The giant tortoises *(Testudo elephantopus)* are found only in the Galapagos Islands and the Seychelles. Their shells can reach a length of 48 inches (1.2 meters) and they are known to live to about 100 years of age. They live in the lowlands but venture to the highlands for drinking water, where they wallow for hours.

The sea turtles (families Cheloniidae and Dermochelyidae) are the most highly adapted to a watery existence. Their legs are flattened and paddle-shaped and they swim with agility. On land, however, they move clumsily and slowly. It is usually only the females that come ashore and they do so specifically to lay eggs in a hole in the sand.

Lizards and snakes

These reptiles belong to the order Squamata, which is divided into two suborders: the Sauria (lizards), and the Serpentes (snakes).

The venomous snakes are grouped into three families according to the position of their fangs. The Elapidae contain front-fanged snakes, whose fangs are fixed, such as cobras and mambas. The Viperidae also contain front-fanged snakes, such as vipers and adders *(Vipera* spp.), but these snakes have long fangs on a rotating maxillary plate which lie against the palate when the mouth is closed. As they open their mouth the fangs swing down and forward. The Colubridae include the rat snakes *(Elaphe* spp.) and are back-fanged snakes. The poison runs down a groove in the teeth at the back of the upper jaw and is injected into the victim when the snake bites it.

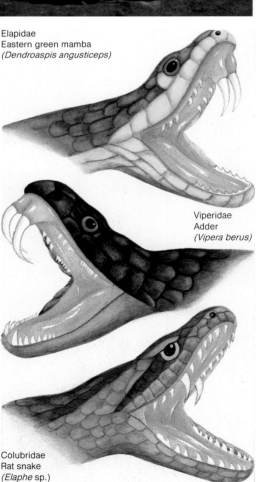

Elapidae
Eastern green mamba
(Dendroaspis angusticeps)

Viperidae
Adder
(Vipera berus)

Colubridae
Rat snake
(Elaphe sp.)

Lizards form the largest group of reptiles, and are most abundant in tropical regions. Most lizards are small and four-legged with long tails. Many can shed their tails to escape from predators. The tail is then regenerated but is never an exact copy of the original, being shorter and with an irregular scale pattern. Members of the family Anguidae have either one or two pairs of short, reduced limbs and some, such as the slowworm *(Anguis fragilis)*, are legless.

An interesting feature of some skinks (family Scincidae) is that the eyelids contain a transparent opening, which allows them to see while they close their eyes to protect them against flying debris.

The family Chamaeleonidae contains some 100 species of chameleons. These lizards are exceptionally well-adapted to an insectivorous and arboreal life. Two of the five toes on each limb are opposed to the other three, forming a gripping "hand," and the tail is prehensile. Each eye can move independently or both can focus on the same object. The chameleon can therefore judge distances accurately and aim precisely at the insects on which it feeds, catching them with its long, sticky tongue.

Only two species of lizards are venomous, with poison glands (in the lower jaw) opening at the base of grooved teeth. These are the Gila monster *(Heloderma suspectum)* and the beaded lizard *(Heloderma horridum)*, both of which live in North America.

The snakes, however, contain most of the poisonous reptilian species, although the majority of species are harmless. Venomous species are classed in three families according to the position and construction of the fangs. The back-fanged snakes such as some African bird snakes *(Thelotornis* spp.) form the family Colubridae. The bird snakes prey on small birds and lizards, which they attract by exposing their brightly colored tongue. The second family (Viperidae) consists of the front-fanged snakes. Members of this family include rattlesnakes *(Crotalus* spp.), the adder, and the asp *(Vipera aspis)*. When their mouth is opened, the fangs swing down and forward. Another group of front-fanged snakes belongs to the family Elapidae, which include the cobras, mambas and kraits, in which the front fangs are fixed. The snakes in the family Boidae, such as the boa constrictor and the pythons (Pythoninae), kill their prey by crushing it.

Unlike lizards, snakes are limbless, although some, such as the boas, have rudimentary hind legs. The tail cannot be regenerated when lost and the transparent eyelids are fused together to form a protective covering. In addition, most snakes are able to dislocate their jaws, enabling them to swallow prey considerably larger than themselves.

Crocodilians

All the members of the order Crocodilia — crocodiles, alligators, caimans and the gharial — are large amphibious reptiles. Crocodilians use their long, vertically-flattened tail for swimming; their webbed feet are usually kept tucked in at the sides of the body. The ears, eyes and nostrils are located high on the head, so that the animal can remain virtually hidden under the water and at the same time can hear, see and breathe. When a crocodilian is totally submerged the eyes are covered by a membrane and the ears and nostrils are closed by special valves. In addition, a fold of skin shuts off the windpipe so that the animal can open its mouth under water without drowning.

Crocodilians are carnivorous and adults prey on large animals — caught either underwater or on land — occasionally as large as a horse. The prey is eaten whole or first torn apart and then eaten. Large prey is often cached under water until it is sufficiently rotted to be torn apart by the powerful teeth; but since their jaws do not move from side to side, crocodiles often resort to twisting their whole bodies in an effort to detach a piece of flesh from a carcass.

The single species, the gharial *(Gavialis gangeticus)* makes up the family Gavialidae. Found only in India, it is distinguished by its long, thin snout and rather bulbous head. The family Crocodylidae comprises all the other members of the order. These include the estuarine crocodile *(Crocodylus porosus)*, found in estuaries and coastal swamps from eastern India to Australia. The American crocodile *(Crocodylus acutus)* occurs in the southeastern United States, south to Ecuador. As in most other crocodilians, the bony plates occur only on the back and tail.

The Mississippi alligator *(Alligator mississippiensis)* and the Chinese alligator *(Alligator sinensis)* are the only two species of alligators. They were once much hunted for their skins and are quite scarce, although in some areas of the United States they are now protected. The South American caimans *(Caiman* spp.) are related to alligators but are quicker moving. They are mainly distinguished by the scutes, which cover both the upper and lower body.

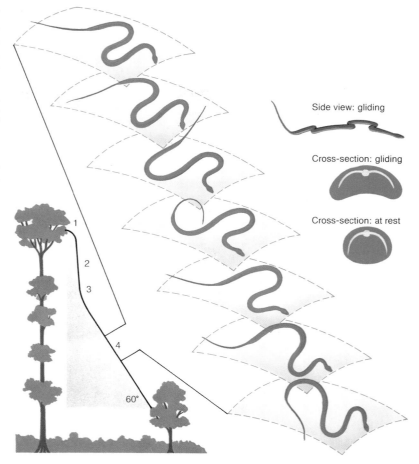

Side view: gliding

Cross-section: gliding

Cross-section: at rest

The tuatara

The tuatara is the only surviving member of the order Rhynchocephalia, a group of reptiles that flourished about 200 million years ago. This lizard-like animal is found only on a few islands off the coast of New Zealand. It preys on insects and other invertebrates and inhabits the burrows of shearwaters and petrels. It becomes active at night and seems able to function at lower temperatures than most reptiles.

The flying snake *(Chrysopelea* sp.) glides by projecting itself from a tree (1), falling (2) for several feet while concaving its lower surface (3), and undulating its body in a series of tight S-curves (4) to gain some lift to slow the descent.

Crocodilian teeth are sharp and cone-shaped. In the crocodiles, a tooth on each side of the lower jaw fits into a pit in the upper jaw. This pit is open on one side and the tooth protrudes when the mouth is closed, giving the characteristic "crocodile smile." One of the most surprising feats, then, is that after the young crocodiles have hatched, the mother carries them gently to safety in her mouth.

Introduction to birds

Birds are relative newcomers to the animal kingdom. The earliest fossil definitely recognizable as a bird, *Archaeopteryx,* lived about 140 million years ago, during the Jurassic Period. It inhabited a world dominated by cold-blooded reptiles, including the pterosaurs, which evolved powered flight at about the same time but became extinct toward the end of the Cretaceous Period, about 70 million years ago.

The fossil record of birds is very incomplete — their fragile bones and feathers have not preserved well. A mere 850 or so fossil species are known, only a tenth of the number of species alive today. After *Archaeopteryx,* of which a few specimens and a single feather have been found, the next fossil birds come from the Cretaceous Period. *Hesperornis* was a waterbird that lost the power of flight. *Ichthyornis* was a flying seabird that probably had teeth, like *Archaeopteryx.* The first representives of the modern families of birds did not appear until the Eocene Period, between 65 and 40 million years ago, and species alive today emerged during the Pliocene, between 13 and 2 million years ago.

The key to the success of birds lay in their development of feathers and flight, and of warm-bloodedness. They have now conquered every habitat, from polar ice to tropical deserts. Some have also adapted to life on the water, conquering three environments — something no other vertebrates have been able to do. Several groups of birds have lost the power of flight and developed large size and strong legs to escape enemies.

The lightest skeleton
The bird skeleton is built on the same basic plan as that of other vertebrates, but it is extensively modified for flight. The whole skeleton has become extremely light. The teeth have been replaced by a lightweight horny bill. Many of the bones, like those of the skull, are very thin, whereas others — such as the limb bones — have a honeycomb structure, being hollow with thin internal struts for strength with rigidity and lightness.

The vertebrae in the backbone near the pelvic girdle are fused together to form the synsacrum, which provides firm support for the legs and cushions the bird against the shock of landing. To make up for this rigidity, the neck is extremely flexible, with many more vertebrae (up to 28) than in the neck of a mammal (with only 7).

Although the basic structure of the limb bones is on the general vertebrate plan, they show significant modifications. The femur (thighbone) is normally hidden because it is held up close to the body beneath the feathers. What looks like the thigh of a bird is really the tibiotarsus, formed from the fused shinbones. The equivalent to our shins is provided by the elongated and fused bones of the ankle and feet, forming the tarsometatarsus. Birds walk on their toes. There are usually four of these, each equipped with a claw. The arrangement of tendons and muscles in a perching bird ensures that when it perches, bending its legs, the toes are automatically pulled inward and the foot grasps the perch tightly — even when the bird is asleep.

The collarbones (clavicles) are fused into a V-shape — the furcula, or wishbone. Together with the coracoids (shoulder blades) they form the pectoral girdle, which stops the ribs being crushed by the powerful wing muscles, which may make up as much as 30 per cent of a bird's weight. The pectoral girdle is attached firmly to the breastbone, or sternum, which is large and keel-shaped, providing a strong anchorage for the wing muscles.

The forelimbs have been modified to form the wings, which articulate with the pectoral girdle. The humerus, the bone of the upper arm, is short and stout and has a large surface area for the attachment of the flight muscles from the sternum. The ulna of the forearm is flattened to accommodate the secondary flight feathers. There are only two wrist bones and three hand bones, two of which are fused together. There are three finger bones; the primary wing feathers are attached to the first of these. A flat membrane of skin on each side of the wing bones, together with the long flight feathers, gives a broad surface to create lift and power.

Unique breathing system
A bird's lungs are connected by tubes to numerous thin-walled air sacs. These can form one-tenth of the volume of the body, spreading into the spaces between the muscles, body organs and

A reconstruction of *Archaeopteryx,* a prehistoric bird which lived about 140 million years ago, reveals that it had teeth and clawed fingers on the front of its wings. Feathers fringed its long tail, and it was probably capable only of short, clumsy flapping flights from tree to tree. Since that time, birds have continued to evolve and adapt, so that today their representatives occupy nearly every terrestrial and semiaquatic habitat on earth. But above all, they have conquered the air.

even the hollow bones. The single-direction air-flow system allows the bird to extract oxygen from the air even at high altitudes where oxygen is in short supply — some birds fly at 25,000 feet (7,625 meters).

Digestive system

The digestive system features a thin-walled, highly extensible crop at the base of the gullet, where food is stored and moistened. The food then starts to be broken down by enzymes in the first part of the stomach. The second part of the stomach is modified to form a thick-walled mus-cular gizzard, which grinds up the food, some-times with the aid of swallowed grit — a bird has no teeth — before it is passed on to the rest of the digestive system.

Both the digestive and reproductive systems open into one chamber, called the cloaca, with a single opening (the vent). Birds rid themselves of waste in the form of solid uric acid — liquid urine would involve losing too much water.

The red-tailed hawk *(Buteo jamaicensis)* is found in North and Central America. Belonging to the buzzard family, the red-tailed hawk is a superb flier and accomplished hunter, and in many ways represents the pinnacle of bird evolution.

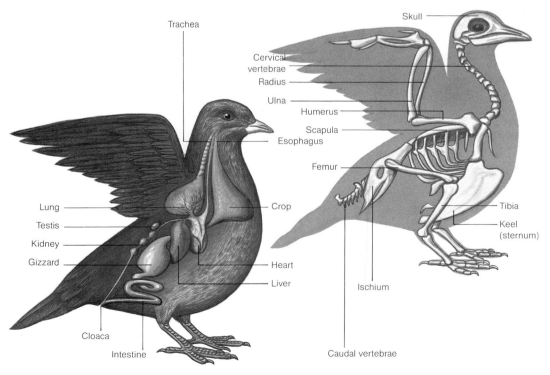

Trachea

Skull

Cervical vertebrae

Radius

Ulna

Humerus

Scapula

Esophagus

Femur

Lung

Testis

Kidney

Gizzard

Crop

Heart

Liver

Tibia

Keel (sternum)

Ischium

Cloaca

Intestine

Caudal vertebrae

The skeleton of a bird is designed for lightness and to provide adequate attach-ment points for the power-ful flight muscles. The main supporting structure is fairly rigid, brought about by fusion of some of the vertebrae in the spine. Nevertheless the whole skeleton is extremely light, because nearly all the bones are hollow. The internal organs follow the broad vertebrate plan, except that in many birds the digestive tract includes a crop in which hard food is ground or predigested. The urinary tracts (from the kidneys) and the lower intestinal tract come together in the cloaca.

Classification of birds

There are more than 8,600 species of birds alive today, compared with only about 4,000 species of mammals. Apart from the nocturnal species, most are fairly easy to observe. Probably for this reason, bird watching is an increasingly popular pastime. As a guide to the enormous variety in the world of birds, this article outlines the makeup of each of the living orders — although not all taxonomists agree on the exact family groupings that constitute separate orders.

The orders of birds

Birds (class Aves) are classified into two subclasses, Archeornithes (containing the extinct *Archeopteryx)* and Neornithes (containing all other birds). The Neornithes consists of four superorders: Odontognathae, extinct toothed birds; Ichthyornithes, also extinct; Neognathae and Impennes containing all other birds, generally arranged in 28 orders (including some extinct orders), subdivided into 158 families. These orders are described in the summaries that follow. The classification is based only on anatomy, behavior and lifehistory. Recently, new techniques of analyzing and comparing proteins (albumins) in egg white and examining parasites have led to changes in groupings.

Order Struthioniformes. The ostrich, the largest living bird, is the single species in this order. It is a large, fast-running, flightless bird, found today in the wild only in Africa. Several other groups of birds have also evolved a flightless life style, an example of convergent evolution. All have breastbones that have lost the keel, because there is no longer any need for attachment of flight muscles. Together with the ostrich, they are known as the ratites (from the Latin *ratis*, a raft, referring to the flat breastbone), separated from all other birds, which are called carinates (from the Latin *carina*, a keel). Their wings are very small, and the feathers loose and fluffy like the down feathers of young birds. They have strong, stout legs for fast running.

Other ratites include the emu of Australia and the cassowaries of Australia and New Guinea (all in the order **Casuariiformes**); the kiwis (order **Apterygiformes**) of New Zealand; and the rheas (order **Rheiformes**) of South America. The two extinct orders, the moas (order **Dinornithiformes**) of New Zealand and the elephant birds (order **Aepyornithiformes**) of Madagascar were also ratites.

Order Tinamiformes. The tinamous are a group of about 50 species of beautifully camouflaged birds that superficially resemble game birds but are probably related to the rheas. They are restricted to South and Central America.

Order Sphenisciformes. The penguins are found only in the Southern Hemisphere, ranging from near the South Pole to the Galapagos Islands on the Equator. Like the ratites, they have abandoned flight. Superbly adapted for fast underwater swimming, their wings have become stiff flippers, their bodies are cigar-shaped and their feet webbed. They eat fishes, squids and crustaceans and generally breed in huge colonies.

Order Gaviiformes. The divers or loons are a primitive group of five species of waterbirds found

Struthioniformes
Ostrich
Struthio camelus

Rheiformes
Common rhea
Rhea americana

Apterygiformes
Brown kiwi
Apteryx australis

Casuariiformes
Emu
Dromaius novaehollandiae

Tinamiformes
Crested tinamou
Eudromia elegans

Sphenisciformes
Magellan penguin
Spheniscus magellanicus

Gaviiformes Great northern diver
Gavia immer

Procellariiformes
Lysan albatross
Diomedea immutabilis

Pelecaniformes
White pelican
Pelecanus onocrotalus

Anseriformes
Black-necked swan
Cygnus melanocoryphus

Falconiformes
Golden eagle
Aquila chrysaetos

Ciconiiformes
Blue heron
Ardea herodias

only in the colder parts of the Northern Hemisphere. Their streamlined shape, paddle-like feet, and long necks and bills equip them superbly for catching fish underwater. The grebes (sometimes given their own order, **Podicipediformes,** but more usually grouped with loons) are equally adapted to an aquatic life; one species has lost the power of flight completely. They are also excellent divers, and are found worldwide.

Order Procellariiformes. This order consists of oceanic species with long tubular nostrils. They seldom come ashore except to breed, and all have webbed feet. Their hooked, plated bills are adapted to deal with a diet of squids, fishes and other marine animals. Best known are the large albatrosses, among the most impressive fliers in the bird kingdom. Other families are the shearwaters, petrels and stormy petrels; and the diving petrels, which have come to resemble the northern auks closely by a process of convergent evolution.

Order Pelecaniformes. A group of large, fish-eating waterbirds, they are the only birds with all four toes webbed. The pelicans are recognizable by their huge bills and highly extensible throat pouches for scooping fish out of the water. The other families are the gannets, or boobies, which dive vertically into the sea for fish from a height of 100feet (30meters) or, more; the tropic birds, graceful seabirds with long tail streamers; the 30 or so species of cormorants and shags, long-necked, long-billed mainly black diving birds; the anhingas, or darters, which have long, snakelike necks and bodies; and the frigatebirds, soarers that frequently pirate food from other seabirds.

Order Ciconiiformes. This order consists of large, long-legged wading birds. The 50 or so species of herons, egrets and bitterns have dagger-like bills for spearing fish and other prey, as do the **16** species of storks. The 30 species of ibises have slender downcurved bills with sensitive tips for probing in mud, whereas the 6 species of spoon-

bills have spoon-shaped bills, which they hold open to trap small animals. The four species of flamingoes live in the tropics, on shallow soda lakes and saline lagoons. Their remarkable hooked bills are equipped with comblike plates that strain out tiny organisms from the water pumped through them by the fleshy tongue; the birds hold their heads upside-down in the water to feed.

Order Anseriformes. The swans, geese and ducks are included in this order, which contains more than 150 species. Predominantly aquatic, they range in size from the mute swan, at 33 pounds (15 kilograms) one of the heaviest flying birds, to the tiny ringed teal, weighing only 10$\frac{1}{2}$ ounces (300 grams). The feet are webbed and the bill typically broad and flattened, with fine plates at the edges for straining food from the water. They breed on every continent and major island except for Antarctica.

Order Falconiformes. This order comprises the birds of prey. A large order with almost 300 species, it includes the carrion-eating vultures and condors. All Falconiformes have powerful, sharp, hooked beaks for tearing flesh, and strong feet armed with sharp talons with which they catch their prey. They hunt by day, using their keen eyesight. Some, like the vultures, kites, typical eagles and buzzards, use soaring or slow flapping flight to spot their prey on the ground. Others, like the falcons, are fast fliers, catching birds and insects on the wing. Still others are specialized for a diet of fish, snakes, snails and even fruit.

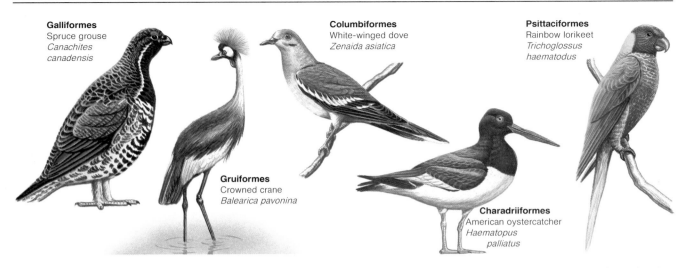

Galliformes
Spruce grouse
*Canachites
canadensis*

Gruiformes
Crowned crane
Balearica pavonina

Columbiformes
White-winged dove
Zenaida asiatica

Charadriiformes
American oystercatcher
*Haematopus
palliatus*

Psittaciformes
Rainbow lorikeet
*Trichoglossus
haematodus*

Order Galliformes. Game birds belong to this order. The 10 species of megapodes, confined to Australasia, are medium-sized brownish birds that build huge mounds of soil and vegetation in which they place their eggs to incubate in the heat produced by fermentation. The 18 species of grouse inhabit the temperate and Arctic regions of the New and Old Worlds. Medium-sized to large, they are plump with short bills and legs. There are some 35 species of pheasants, including the red junglefowl of south-eastern Asia — the ancestor of the domestic hen — the peacock and the smaller partridges and quails. The seven species of guinea fowl live in open country in Africa; and the two species of turkeys are found in the woodlands of North and Central America.

Order Gruiformes. These are ground-nesting, ground-feeding birds. Many are poor fliers, although the migratory cranes are a striking exception. Large, long-legged, grey, white and black or brown birds, the 15 species of cranes have long secondary wing feathers drooping down over the tail; some have ornamental crowns or tufts of feathers on their heads, used in their spectacular mating dances. They inhabit Eurasia, North America and Australia. The rails are among the most widespread of all land birds, found on all continents and very successful at colonizing remote oceanic islands. Most island rails have become flightless. They range in size from species little larger than a sparrow to birds as big as a duck. There are about 130 species of rails. They have short, rounded wings, small tails and long legs, and include the aquatic coots and gallinules.

Order Charadriiformes. This order consists of the waders, gulls and their relatives, typically found on or near seacoasts and freshwater. The waders are a huge group of over 200 species, ranging from the crow-sized oystercatchers and larger curlews to the medium-sized plovers and diminutive stints. Most are long-legged, and many have long bills for probing for food in mud. The skuas are dark plumaged gull-like seabirds with webbed feet and most have elongated central tail feathers. They feed on fish, small mammals, birds and their eggs, and also chase gulls, terns and other birds, forcing them to disgorge their food. There are five

or six species. The gulls are a successful and wide-ranging group. There are 43 species, from the large great black-backed gull — 30 inches (75 centimeters) — to the little gull — 10 inches (25 centimeters). They are typically brown when immature and grey-and-white or black-and-white as adults. Most gulls are coastal, but some live inland; many have expanded their range dramatically, benefiting from food thrown away by man. Terns are slender white, grey-and-white, black or black-and-white seabirds with short legs, webbed feet and long pointed bills; they are found worldwide. The auks are short-winged black-and-white diving birds restricted to the northern oceans. Although they can fly, they are most at home in the water, and are the northern equivalent of the penguins, also nesting in huge colonies. Size ranges from the tiny little auk — 6.5 inches (16 centimeters) — to the extinct great auk — 30 inches (75 centimeters). The 21 living species include the familiar puffins.

Order Columbiformes. This order contains the pigeons and their allies, including the extinct dodo. The pigeons and doves (300 species) are found throughout the world, except for Antarctica. They live on seeds, fruits and berries. They are able to drink by a suction action without lifting their heads from the water.

Order Psittaciformes. These are the parrots, the 315 species of which are mainly restricted to the tropics. Varying in size from 4 inches (10 centimeters) to 40 inches (1 meter) long, most are mainly green, although some are brilliantly colored and many have long tails and stout, short strongly-hooked bills adapted for dealing with fruits, berries and seeds.

Order Cuculiformes. This order includes the 18 species of brightly colored turacos from the jungles of Africa and the cuckoos, some of which are parasitic breeders. They lay their eggs in the nests of other birds and allow them to rear their young. More than 100 species of cuckoo are distributed worldwide.

Order Strigiformes. Most of the 131 species of owls are nocturnal birds of prey, with large eyes at the front of their heads, hooked flesh-tearing bills, razor-sharp talons and dense, soft plumage

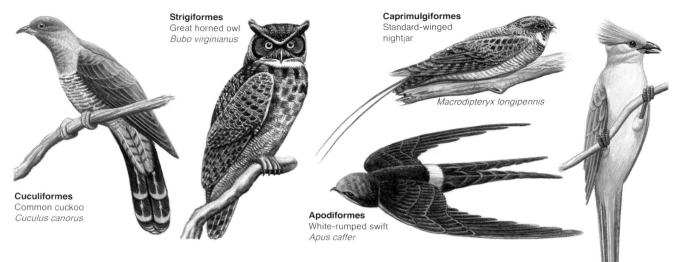

Strigiformes
Great horned owl
Bubo virginianus

Caprimulgiformes
Standard-winged
nightjar
Macrodipteryx longipennis

Cuculiformes
Common cuckoo
Cuculus canorus

Apodiformes
White-rumped swift
Apus caffer

Coliiformes
Blue-naped mousebird
Colius macrourus

making them almost noiseless in flight. They are subdivided into the 10 species of barn and bay owls (family Tytonidae) and the 121 species of typical owls (Strigidae). Owls are found throughout the world, except for Antarctica.

Order Caprimulgiformes. This order contains about 95 species of nightjars (sometimes called goatsuckers) and their relatives, found worldwide except for Antarctica. Nocturnal birds related to owls, they have long, pointed wings and bills fringed with bristles for trapping flying insects.

Order Apodiformes. All the members of this order are exceptional fliers with pointed, slender but powerful wings and tiny feet that are useless for walking. The swifts are the most aerial of all birds, able to fly for days on end, even sleeping on the wing. The 75 or so species are found virtually throughout the world. The 320 species of hummingbirds are birds of the New World, which can hover and even fly backwards. Most are tiny and the largest is only the size of a sparrow. They feed on nectar and insects.

Order Coliiformes. This order consists of the single family of coly or mousebirds. All six species are African and are small birds with very long, stiff tails and prominent crests.

Order Trogoniformes. This order contains the brightly colored tree-dwelling tropical trogons,

which have unusually delicate skins. Like the mousebirds, their relationship to other birds is obscure.

Order Coraciiformes. The brightly colored, mainly tropical birds of this order nest in holes in banks or trees, and have the front three toes partly joined. They include the 87 species of kingfishers, 8 species of motmots, 24 species of bee-eaters, 16 species of rollers, and 8 species of wood hoopoes and hoopoes, as well as the 44 species of huge-billed hornbills.

Order Piciformes. Birds in this order have feet with two toes pointing forward and two backward, and include the 208 species of woodpeckers, specialized for feeding and nesting in tree-trunks, and the giant-billed fruit-eating toucans (37 species) of the tropical rainforests of Central and South America.

Order Passeriformes. This order, the perching birds, is the biggest group of all, containing more than a third of all living families and over half the living species. All have feet adapted to perching on or clinging to branches or other supports. The order includes the "song birds" which have developed the ability to sing to the highest degree. It also contains the swallows, wagtails and pipits, wrens, thrushes, warblers, tits, finches, weavers, and sparrows, starlings and crows.

Trogoniformes
White-tailed trogon
Trogon viridis

Coraciiformes
Common kingfisher
Alcedo atthis

Piciformes
Ivory-billed
woodpecker
*Campephilus
principalis*

Passeriformes
Wood thrush
*Hylocichla
mustelina*

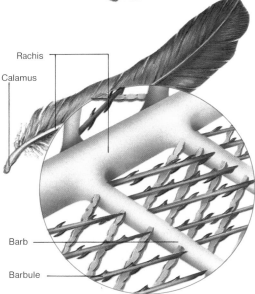

Bird flight is complex, but the main feature is the wingbeat. The sequence (above, right) begins with an upstroke, in which the primary feathers open to allow air to flow through them (A to D). The wing has such a high angle of attack on the downstroke (E to G) that it would stall but for the separation of the primaries, which act like individual wings. A flight feather (right) has a central midrib (rachis) bearing a flattened vane made up of hundreds of interlocking barbs, each of which bears tiny barbules that hook together. The calamus, at the base of the rachis, is hollow for the passage of nutrients to the feather.

Rachis

Calamus

Barb

Barbule

Physical characteristics of birds

The two factors that set birds as a class apart from nearly all other vertebrates are feathers and flight (only bats, which are mammals, are also capable of powered flight). Feathers provide insulation and the lifting surface of the wings, and the mobility provided by flight has enabled birds to spread and establish themselves in nearly every habitat on earth.

Feathers
Evolved from the scales of their reptilian ancestors, bird feathers are made of the same basic material — the protein keratin — as their horny bills and the scales on their legs. Each feather develops from a knob (papilla) within a feather socket, or follicle. The follicles are arranged in distinct areas called feather tracts over the bird's body. Each follicle produces one, two or even three sets of feathers every year.

There are two main types of feathers — pennae and plumulae. The pennae are the flight and contour feathers. The flight feathers (primaries and secondaries) are directly concerned with flight, whereas the contour feathers have several functions. They give the bird its streamlined shape, helping it to cut through the air efficiently. They also provide vital insulation; heat is retained by the layer of air trapped close to the skin by the feathers. During hot weather, a bird opens up its contour feathers to allow heat to escape, and in cold weather it fluffs them up to increase the insulating layer of air.

Another function of feathers is to waterproof the bird. They are also the main source of the bird's color, whether for camouflage or for species recognition, sexual display or as warning signals.

The plumulae are the down feathers. They lie beneath the contour feathers, providing extra insulation, and are the only feathers on a newly hatched chick. All other feathers are intermediate between or derived from the two basic types. Down feathers are much simpler than pennae, with a very short midrib and no barbules, so that the barbs are separate, giving them their fluffiness.

The number of feathers varies considerably from one species to another; usually, the larger the bird, the more feathers it has. For example, the tiny hummingbirds may have fewer than 1,000 whereas a large swan may have more than 25,000. The number of feathers also differs from one season to another.

Feathers are subject to great stress, and wear out. All birds molt at least once a year, many twice and some even three times. Most birds molt their feathers gradually, so there is no interference with flight; but ducks, for example, lose all their flight feathers at once (during an eclipse period), when they are flightless and vulnerable.

Despite their proverbial lightness, all the feathers on a bird are, surprisingly, often heavier than its incredibly light, hollow-boned skeleton. In the bald eagle *(Haliaeetus leucocephalus)*, for instance, which weighs about 9 pounds (4,000 grams), the feathers account for more than 1.5 pounds (670 grams) but the skeleton for only about 6 ounces (270 grams).

The shapes of birds' bills reflect adaptations for dealing with different types of food. The strong hooked bills of cormorants and eagles tear the flesh of their prey, whereas those of macaws and toucans crack nuts and deal with fruit. A gull's "all-purpose" bill suits its role as a scavenger, and a duck uses its flattened bill to strain food from the water. A blue tit's short bill catches insects and grubs, whereas as the extremely long narrow bill of the hummingbird allows it to reach the nectar deep inside tubular flowers.

C   D   E   F   G

Flight

Bird flight is governed by the same laws of aerodynamics that control all heavier-than-air flight. There are two opposing forces involved: lift, the upward force that keeps the bird airborne, and drag, the force of the air opposing lift, which is to a great extent the result of air turbulence at the wingtips.

A bird's wing is curved in cross-section, rounded above and hollow below, and with the leading edge thicker than the trailing edge — a shape that produces maximum lift. The convex upper surface causes the air flowing over it to travel faster than the air flowing past the undersurface. As a result, the pressure on the underside of the wing is greater than that on the upper surface, producing lift.

The angle the wing makes with the horizontal is called the angle of attack. There is an optimum angle for flight, at which upward lift counteracts the forces of gravity and of drag. The greater the angle of attack, the more the airflow over the wing becomes turbulent, and the more the bird is likely to stall.

Powered bird flight is very complex, and not yet fully understood. The highly flexible primary feathers of the wingtips twist, acting almost like a propeller to produce the forward thrust that drives the bird through the air. The wingtips move faster than the rest of the wing. The main function of the secondary feathers is to provide lift; they move but little as the wing beats. Slow-flying birds, such as herons and eagles, are often in danger of stalling and have large wing slots formed by the separation of the alula, or bastard wing, at the front of the wing. The frequency of wingbeats varies greatly, from about two beats per second in a swan to 50 or more per second in a hummingbird.

To alter the direction of flight, a bird increases

the angle of attack of one wing, or beats that wing faster, creating a difference in lift on the two sides of its body. Movements of the tail are also used. To land as gently as possible and minimize the shock to its body, a bird must be on the point of stalling. To achieve this, it pushes its body downward and spreads its wings and tail.

Bird flight-speeds also vary, although typical speeds are between 9 and 60 miles (15 and 95 kilometers) per hour. The fastest bird is probably the peregrine falcon *(Falco peregrinus)*, which dives on its prey at speeds of 180 miles (290 kilometers) per hour or more.

Some birds save energy by gliding instead of flapping their wings. Vultures, for instance, use their long, broad wings to soar in thermals (rising currents of warm air). Albatrosses, with long, narrow wings, tack back and forth low over the waves, using the updrafts produced by the friction of the air with the water to stay aloft.

Hovering in midair, a tiny ruby-throated hummingbird *(Archilocus colubris)* uses its long bill to feed on nectar from sage flowers. The wings beat at more than 70 times a second, requiring its heart to beat 600 times a minute and using five times as much oxygen (per ounce of body weight) than a larger, slow-flying bird.

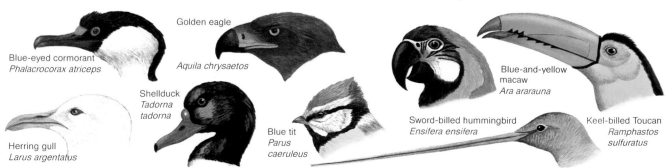

Blue-eyed cormorant
Phalacrocorax atriceps

Golden eagle

Aquila chrysaetos

Shellduck
Tadorna tadorna

Herring gull
Larus argentatus

Blue tit
Parus caeruleus

Blue-and-yellow macaw
Ara ararauna

Sword-billed hummingbird
Ensifera ensifera

Keel-billed Toucan
Ramphastos sulfuratus

Breeding and behavior in birds

Most birds breed within a definite territory, an area that they defend against rivals. In many perching birds, it is about an acre in extent; but in large birds of prey — which obtain food far less frequently over a much wider area — it may be as much as 30 square miles (75 square kilometers) in area. Territory helps birds to survive by parceling out the available habitat into areas capable of supporting a pair of birds, allowing them to feed and breed there.

Courtship

Birds recognize potential mates of the same species and breed with them following courtship. A highly ritualized and often elaborate form of behavior, it involves complicated visual displays or calls which help to establish and reinforce the pairbond between the sexes. The type of courtship behavior varies, as does the type of pairbond formed. Many birds are monogamous and establish lasting bonds, but others are polygamous, successful dominant males mating with several females.

A vital part of bird displays is the emphasis on the colors and patterns that distinguish a particular bird from otherwise similar species. Where several closely related species coexist, the plumage of the males tends to be distinctive, differing strikingly between species (as with ducks of the Northern Hemisphere). On the other hand, in areas with only one species (as with single species of finches on an oceanic island) the plumage can be quite nondescript. Where birds of different species that look very similar do coexist, as they do with the (to us) virtually indistinguishable willow warbler *(Phylloscpus trochilus)* and chiffchaff *(P. collybita)* of Eurasian woodlands, their songs identify the birds.

Another important function of courtship is to encourage the male and female to reduce territorial hostility and accept each other as mates — a process that may take several weeks. For example,

the male European robin *(Erithacus rubecula)* sings at the boundaries of his territory, both to warn off rival males and attract potential mates. Sooner or later, his song encourages a female to approach him. The male first responds by adopting a threat display, using his red breast as a "flag" to discourage the intruder, just as he would if a rival male was involved. A rival male would either return the threat or flee. But the female does neither, adopting a submissive posture instead. This gradually encourages the male to accept her in his territory, and eventually to mate, nest and rear a family.

Unlike the robin, some male birds do not display in an individual territory, but come together at communal display grounds, or leks. Gamebirds such as the American grouse *(Centrocerus urophasianus)* or the capercaillie *(Tetrao urogallus)* of northern Eurasian coniferous forests, certain waders such as ruffs *(Philomachus pugnax)*, and birds of paradise *(Paradisaea* sp.), all have leks. The females go there only to pair and mate and take no part in the display, leaving this entirely to the gaudy males.

Not all birds have displays in which the male performs or sings and the female takes on a submissive role. Many species have courtship rituals that involve both partners equally — examples are the strange, complex dances of grebes *(Podiceps* sp.), the elaborate bill-scissoring displays of some albatrosses *(Diomedea* sp.), and the wild dances of many cranes *(Grus* sp.).

Displacement activities

Detailed study of courtship ceremonies has shown that they are usually modified and highly stylized versions of everyday actions, such as preparing for flight, preening, drinking or feeding. Courtship is a tense period, involving the conflicting emotions of aggression, fear and the sex drive, and so a bird may indulge in ritualized behavior which alternates between a desire to flee and a desire to approach the mate — the "fight or flight" situation.

If the conflict is great enough, as when a female behaves aggressively towards a male, the bird may show "irrelevant" behavior, such as beak-wiping; this is called displacement activity. In many birds, displacement activities have become ritualized and form an important part of their courtship display. Other courtship ceremonies originate from the behavior of young birds; for example, female finches and many other birds flutter their wings and beg food from the male before mating.

Mating

Successful copulation, resulting in fertilization of the egg or eggs, depends on the full cooperation of the female; unlike mammals, most male birds do not have a penis. Sperm is transferred from the cloaca of the male to that of his mate. To encourage better contact between them, the female sticks out her cloaca, as well as moving her tail to one side.

Mating takes place mainly during the time when the female is carrying unfertilized eggs. The single egg of the Emperor penguin *(Aptenodytes fosteri)* needs only one fertilization, whereas birds such as songbirds, which lay several eggs, may involve several fertilizations over a period of time.

Nests and nesting

Birds vary enormously in their choice of nest sites, nesting materials and methods of nest construction. Some penguins and the auks, for instance, build no nest at all. The Emperor penguin incubates its single egg on its feet — where it is warmed by a feathered flap of belly skin — for nine weeks of subzero temperatures and perpetual darkness during the Antarctic winter. Most waders nest on the ground, generally providing

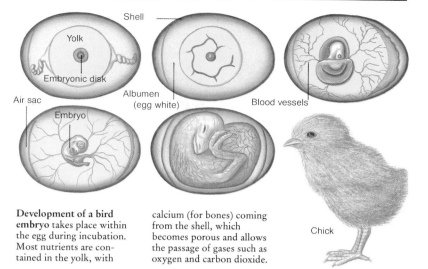

Development of a bird embryo takes place within the egg during incubation. Most nutrients are contained in the yolk, with calcium (for bones) coming from the shell, which becomes porous and allows the passage of gases such as oxygen and carbon dioxide.

little more than a few scraps of vegetation or pebbles for their eggs. Some species of gannets, auks and gulls nest on virtually inaccessible seacliffs where they are safe from mammal predators. The eggs of guillemots (auks), which nest on particularly perilous ledges, are pear-shaped and as a result they roll around in circles instead of over the edge of the cliff.

Many birds, such as owls, woodpeckers and hornbills, nest in holes in trees, and kingfishers, some swallows and bee-eaters nest in holes in sandy banks. Such birds generally lay white eggs, since there is no need for camouflage in these situations. Others, such as herons, crows and many birds of prey, nest high up in tall trees, building strong, bulky nests of branches and large twigs. Some of the cave swiftlets *(Collocalia* sp.) of Asia build nests of their saliva alone, which they glue to the walls or roofs of caves; these are the nests collected for making birds' nest soup.

Contrasts in nest-building are demonstrated by herons and weaverbirds. The great blue heron *(Ardea herodias,* left) lives in Canada and the United States and builds ragged nests of twigs in colonies perched on tall trees and buildings. The black swamp weaver *(Amblyospiza albifrons,* right) builds a delicate nest in reeds by interlacing lengths of grass to form a globular home with an entrance hole near the top.

Camouflaged coloration is a significant feature of ground-nesting birds and their eggs. During the breeding season, the speckled back plumage of the golden plover *(Pluvialis apricaria)* makes it almost invisible.

A small dunnock, or hedge sparrow *(Prunella modularis),* tries to keep up with the voracious appetite of its enormous "offspring," a cuckoo *(Cuculus canorus)* which has tipped the dunnock's own eggs out of the nest.

Eggs vary in shape from the almost spherical eggs of owls and kingfishers to the long, pointed eggs of swifts, waders and guillemots. They also vary enormously in size, from the tiny eggs of hummingbirds (as small as half an inch long and only 0.2 ounce in weight) to the huge ones of ostriches — 8 inches (20 centimeters) and 3.3 pounds (1.5 kilograms).

Unlike those of fishes and amphibians, the eggs of birds are adapted to allow the embryo to develop on dry land. A bird's egg is a closed system. The large yolk feeds the growing embryo and there is built-in protection in the form of the tough shell which, nevertheless, is porous, allowing air to pass in for the embryo to breathe and letting out excess water. The chick excretes uric acid, which is insoluble and does not dissolve in the body fluids and cause harm; this waste is secreted into the baglike allantois, which is left behind in the shell after hatching.

Incubation
Unless the eggs are kept at a relatively high temperature — about 102°F (39°C), the chicks will not survive. Responding to hormones, most birds develop naked brood-patches on their abdomen, well supplied with blood vessels, to transfer their body heat to the eggs. In many birds, both sexes share the task of incubation. In a few, such as cassowaries and the emu, only the male incubates, whereas in others, like many game birds and the owls, the female sits on the eggs. The Australasian megapodes (brush turkeys, *Leipoa* sp.) do not brood; instead, they build huge mounds of decomposing vegetation which generates enough heat to incubate and hatch the eggs.

Hatching
The chick breaks out of its protective prison by chipping out a series of holes in the middle of the shell with an egg-tooth at the end of its upper bill, then cracking it open by pushing in opposite directions with its head and feet.

Chicks are of two basic types. Nudifugous ("nest-fleeing"), or precocial, chicks leave the nest within a few days of hatching; some, such as the chicks of plovers, find all their own food from the start. They are well equipped with a covering of warm down feathers, their eyes are open, and they are usually superbly camouflaged against predators. Examples of this first type are the young of wild fowl, game birds and waders. The second type, nestlings, are known as nidicolous ("nest-inhabiting"), or altricial. They hatch at a much more immature stage of development, often naked, blind and helpless. They remain in the nest until almost ready to fly and are totally dependent on their parents. Examples are the young of pigeons and owls, or of warblers, thrushes, tits and other passerines.

Migration
One of the main reasons birds are so successful is that they are capable of moving far and fast. Many species make long migrations in search of food or

The role of the sexes in nest-building varies greatly. In most birds, both male and female are involved, but in finches and hummingbirds, among others, the female does most of the work. In some, such as weaverbirds, the males are the nest-builders.

Eggs and egg-laying
After the male's sperm has fertilized the egg in the female's ovary, the egg — now consisting of an ovum (egg cell) plus the yolk — moves down the oviduct. It is covered firstly by the jelly-like "egg-white," or albumen, then by two shell membranes and finally by the shell itself, which consists of several layers covered by a thin cuticle. The pigments that give the egg its color are laid down mainly in the cuticle and outer shell layers. Mixtures of just two basic pigments, red-brown and blue-green, give eggs their great range of colors and patterns. White eggs contain no pigments.

The farthest-traveling migrant bird, the Arctic tern *(Sterna paradisaea),* flies annually up to 22,500 miles (36,000 kilometers) from its breeding grounds in the north to the Antarctic and back again. For much of the year it lives over the oceans, flying along the eastern shore of the Pacific Ocean or across the Atlantic. At the most northerly and southerly extremes of its range, the bird lives in perpetual daylight.

Breeding range

Non-breeding range

breeding sites. True migration involves regular seasonal journeys between a breeding area and a wintering area where the climate is more favorable and the birds can find food. Many insect-eating birds, such as swallows and warblers, breed in temperate regions and migrate to lower latitudes in autumn, to spend winter in the tropics. Others, including some wild fowl, waders and some thrushes, breed in high latitudes and migrate to temperate regions for the winter.

Navigation

Many birds follow well-defined migratory flyways, which often parallel coastlines, mountain ranges and valleys, or ocean currents. It is clear that such birds may follow geographical clues; they do not need to be taught their route. Young Eurasian cuckoos *(Cuculus canorus),* for instance, find their way back thousands of miles to their winter quarters in Africa completely unaided, their parents having returned several weeks before them. But the instinctive following of landmarks cannot explain the phenomenal journeys of seabirds that fly right across the vast Pacific Ocean, where there are no clues to guide them. Nor can they explain the fact that a Manx shearwater *(Puffinus puffinus)* taken from the coast of Wales and flown in an aircraft to Boston in the United States, found its way back to its nest over more than 3,000 miles (4,900 kilometers) of open ocean in $12\frac{1}{2}$ days — arriving before the letter announcing its release.

Despite many years of research, much remains to be learned about how birds perform such astonishing feats of navigation. According to some experts, it is possible that they possess a sort of internal compass which enables them to use the Sun by day and the stars by night to get their bearings. It also seems likely that they are able to respond to the Earth's magnetic field and navigate by taking magnetic bearings, especially when the sky is overcast.

Migrating geese fill the sky as they travel southward from their breeding grounds in Arctic Canada. Despite the apparent confusion as they take off together, the geese soon group themselves into a characteristic V-shaped formation (see inset). In autumn the white snow goose *(Anser caerulescens)* and its smaller blue subspecies *(A. caerulescens caerulescens),* both of which appear in the larger photograph, travel 1,700 miles (2,700 kilometers) at an average of nearly 30 miles (50 kilometers) per hour to winter in the marshes fringing the Gulf of Mexico.

Introduction to mammals

The class Mammalia is a relatively small one, especially when compared to the insects (it contains some 4,000 species, whereas Insecta comprises nearly a million). The mammals contain 19 orders, which include rodents (Rodentia) — which make up about half of the mammals — and bats (Chiroptera), which account for about one quarter of mammals. Mammals are one of the most diverse groups, varying remarkably in structure and size, and in the habitats in which they are found. The smallest living mammals are the shrews (family Soricidae); the pygmy shrew *(Microsorex hoyi)*, for example, weighs less than an ounce. The largest mammal is the blue whale *(Sibbaldus musculus)*, with a weight of up to 145 tons.

Distinguishing features
The main distinguishing features of mammals are the presence, in the female, of mammary glands (mammae), a body covered in hair, warm-bloodedness, and a large brain.

Mammary glands are found in all species without exception, although there is some variation in form among primitive types, such as monotremes and marsupials. The mammae produce milk for suckling the young and provide them with nourishing food. They also help to maintain the tie between mother and newborn for a considerable period. This dependence of the young on the mother means that there is a period during which it can learn from the experience of its mother, and so become more efficient in dealing with the problems of survival.

In most mammals, hair or fur covers the entire body. Its purpose is to act as an insulator, because mammals are warm-blooded and need a means of preventing the loss of body heat. Body hair is one factor in enabling a mammal to maintain a fairly constant temperature — normally between 96.8°F (36°C) and 102°F (39°C). But in addition, a mechanism (the hypothalamus) situated in the brain regulates body temperature. In most mammals this mechanism works together with sweat glands which are found in the skin and distributed over much of the body. (Dogs have them only in their feet and keep cool mainly by panting.) Heat induces the hypothalamus to become active which causes blood vessels to dilate and give off the heat of the blood. The sweat glands exude a liquid onto the skin so that it is cooled by evaporation. When temperatures are low, the blood vessels contract and the sweat glands dry up. In addition, reflex shivering occurs which generates heat by the successive contractions of muscles.

Because the body temperature is kept fairly constant, mammals can usually remain active regardless of the external temperature, although in extreme conditions of cold or heat, some may hibernate, or estivate (sleep during the dry season). The body becomes torpid as the rate of heart beat, breathing and other body processes slow down considerably. The animal then seeks shelter in a burrow or other safe place until the outside temperature increases.

Mammals have a four-chambered heart (also found in fishes, birds and crocodiles), and a diaphragm which separates the chest and abdominal cavities so that the lungs can work more efficiently and therefore increase the amount of oxygen supplied to the blood. Most mammals have fleshy lips that are initially used in suckling, but as the animal grows, the mouth may become adapted to feed on specialized food.

Habitats
Mammals are regarded as extremely successful animals. One of the reasons for their success is that they are highly adaptable and as such have been able to exploit an amazing variety of habitats. Some make their homes in underground burrows, such as moles (family Talpidae) and many rodents, or among leaf litter on the ground (as do shrews); others lead a semiaquatic life in freshwater, such as otters (family Mustelidae) and beavers (family Castoridae), or spend all of their

Suckling piglets typify young mammals, all of which are initially nourished by mother's milk. Unlike most mammals, however, adult domestic hogs have hardly any fur or hair, just a sparse scattering of bristles, although this feature is shown by the fine velvety hair on the piglets. The hog *(Sus scrofa)*, bred from the wild hog, often has large litters of between 6 and 14 piglets.

life in the sea such as whales do (order Cetacea). Mammals also inhabit the tree canopies in forests (monkeys and others) and some live in the air at least part of the time, such as bats. Certain mammals are able to live in Arctic conditions without having to hibernate, as do Musk oxen *(Ovibos moschatus)*, whereas others live successfully in desert conditions and can withstand the full heat of the sun, such as camels *(Camelus* spp.).

Dentition

Because mammals are warm-blooded they consume more energy than cold-blooded animals do, and need to increase their food intake to fuel their metabolism. They need teeth in their mouth to break up food at that point so that digestive processes can start earlier and get more out of the food than if the food were processed only in the stomach.

The teeth closely reflect the diet of different mammals. They include incisors, canines, premolars and molars. Incisors are cutting teeth in the front of the jaws, and are extremely well-developed in the rodents, designed for cutting through tough woody materials. Canines, which are used in tearing and piercing flesh, are highly developed in the carnivores, although their incisors are small. These are used for nibbling, play and for grooming. The flat-topped premolars and molars are used for grinding non-woody vegetation and are the most important teeth for herbivores, which graze on grasses and other herbs, or browse on leaves. In carnivores, the last upper premolar and the first lower molar have become bladelike to slice through flesh, and are known as carnassials. These teeth cut meat into bite-sized pieces by their scissor action. Some mammals, such as pangolins and anteaters (which feed on termites and ants) or baleen whales (which feed on shrimplike crustaceans) have no teeth at all and use special mechanisms for collecting and controlling their prey. The giant anteater *(Myrmecophaga tridactyla)* uses its long, sticky tongue

to sweep up insects from the ground. Its digestive system includes a gizzard which grinds up the hard exoskeleton of ants.

Locomotion

Most mammals are quadrupeds — they move about on four feet. Some walk by placing the whole foot on the ground, when they are called plantigrade. This method results in relatively slow locomotion and is seen, for example, in the polar bear *(Thalarctos maritimus).* Many mammals, such as cats and dogs, lengthen their stride by walking on their toes; this is known as digitigrade locomotion and enables the animals to run at high speeds. Other animals walk on the tips of their toes which are normally protected by large nails or hoofs. This is called unguligrade locomotion

The thick white fur of the polar bear *(Thalarctos maritimus)* and its layers of fat insulate it against the bitter Arctic cold. It is able to survive where few mammals do, apart from seals, walruses and whales. Unlike most animals, the polar bear has fur on its flat feet, apart from the toe and foot pads. This fur gives it extra grip when walking on the ice.

The limbs of mammals conform to a basic pentadactylic pattern, sometimes much modified. The lower limb consists of the heavy, weight-bearing tibia and the slighter fibula, and a foot with five toes, containing the ankle bones (tarsals), metatarsals and phalanges. There are three phalanges on each digit except for the large innermost one. The structure of the foot varies among species. Plantigrade feet are found in humans and bears and are constructed so that the whole foot is placed on

the ground. Animals such as dogs and cats, which run fast, have digitigrade feet in which only the part containing the phalanges is used for walking on. Ungulates walk on the tips of the last phalanges. This is called unguligrade locomotion. In horses, the tibia and fibula have fused to form the cannon bone, as have the tarsals to form the splint; the only remaining toe is the third one. Aquatic mammals, such as seals, have adapted the foot to assume a fin-like shape.

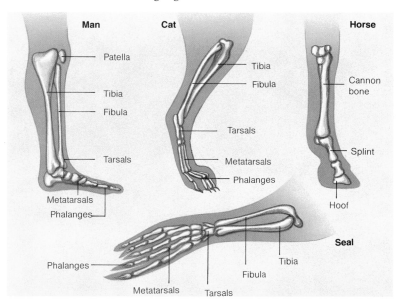

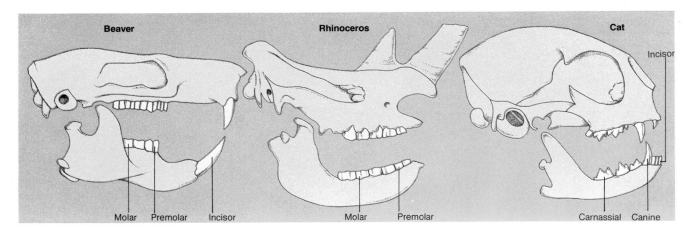

Beaver   **Rhinoceros**   **Cat**

Incisor

Molar   Premolar   Incisor   Molar   Premolar   Carnassial   Canine

The dentition of mammals often reflects their diet. The jaws of herbivorous rodents, such as the mountain beaver *(Aplodontia* sp.), have very large upper and lower incisors to gnaw at woody vegetation, and no canines. Their premolars and molars are almost flat and therefore good for grinding vegetation. Grazing vegetarians such as rhinoceroses (Rhinocerotidae) usually have no incisors or canines. Carnivores, such as cats (Felidae), have small incisors, but canines for tearing at flesh; their premolars and molars (carnassials) have sharp edges which slice through meat.

and is found in deer, antelopes and horses — all of them fast, often long-distance, runners. Humans are the only animals to walk bipedally in an erect position at all times, but some apes and monkeys move in this way for short periods.

Aquatic mammals have limbs which have evolved into paddles, or fins. In some, such as the otter and the platypus *(Ornithorhynchus anatinus)*, a membrane has evolved between the digits on the forelimbs, which aids them in swimming. In seals (Pinnepedia) all the flippers are used for moving through water, whereas in whales, the hind limbs have completely disappeared; locomotion is powered by movements of the back and tail whereas the function of the foreflippers is only to steer and balance.

The gibbons, orangutans and some other arboreal species of the Pongidae family have strong arms which they use to swing from when traveling from branch to branch. Some South American monkeys also have prehensile tails which operate as a fifth limb. The scaly-tailed squirrels (Anomaluridea), which also live in the tree canopy, have membranes which extend from the forelimb to the hindlimb and tail, on which

they glide from tree to tree. They are sometimes called flying squirrels, but the only mammals that are true fliers are the bats. Their forearm is the major support of a wing membrane. The hindlimb and the tail often also support the wing.

Classification of mammals
The class Mammalia is divided into two subclasses: the Prototheria and the Theria. The first is a group of primitive mammals and today contains one order only — the Monotremata. It comprises two families: the Ornithorhynchidae (birdnoses), with one species — the duck-billed platypus; and the Tachyglossidae — the spiny anteaters, or echidnas. Monotremes, like reptiles, are oviparous, or egg-laying, and have low, or variable, body temperature. They are toothless, although young platypuses produce three tiny teeth soon after birth, which they lose and replace with horny plates. Also, like reptiles, monotremes have one opening only at the hind end of the body — the cloaca — which serves both the processes of elimination and reproduction. But unlike reptiles, the females have mammary glands, although they do not have mammary

Elephant seals *(Mirounga* spp.) are the largest of the seal group. The males weigh about three tons and can measure about 20feet (6meters) in length. Their name is derived from the bladder on their nose, which can be 15 inches (38 centimeters) long. These animals are polygamous; the bulls become belligerent in the breeding season and challenge each other's dominance to acquire a harem.

organs. Milk secreted by the mammae lies on the surface of the skin either on the abdomen, or, in the echidnas, in the pouch, from where the young lap it up, rather than suckle. Female platypuses also differ from other mammals in that they have one ovary only, on the left side of the body whereas other mammals have two ovaries, on either side.

The subclass Theria contains all other living mammals which give birth to live young. It is divided into two infraclasses: the Metatheria, which has only one order — the Marsupialia, or pouched mammals; and the Eutheria, or placental mammals. The marsupials give birth to live young but they are born at a very early stage of their development, blind and greatly undeveloped, when they are called neonates. As soon as they emerge, they crawl through the mother's fur to her pouch where they remain attached to the teats for several months before they are weaned.

In the eutherians, or true mammals, the development of the young within the mother is much more complex. Like all mammals, fertilization is internal. Once the egg, or ovum, has been fertilized by sperm it becomes fixed to the wall of the uterus. (Placental mammals have one uterus only, although in some it is divided into two, or is two-horned.) As in birds and reptiles, special structures then develop around the embryo — a protective sac called the amnion which contains amniotic fluid; and the allantois, or respiratory membrane. In addition a placenta is produced from tissues that arise from both mother and embryo. It is usually attached to the uterus wall and connected to the embryo by the umbilical cord. The purpose of the placenta is to allow the passage of oxygen and nutrients from mother to embryo and waste products from embryo to mother. These substances are carried in the blood, although no blood passes between the two. Rather, they pass through a thin partition separating the two blood systems. This form of development is highly efficient and allows the fetus to remain inside the mother until it is very well developed. Even so, in most mammals the newborn are not fully mobile and are often helpless, as in bats and most rodents.

In most males two scrotal sacs are found which

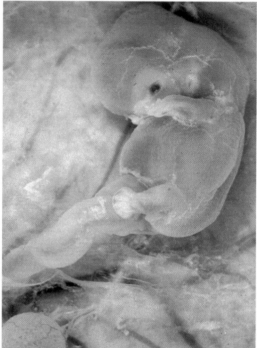

protrude on the ventral surface of the body just in front of the pelvis. They fuse behind the penis. In marsupials, however, they lie in front of the penis. In many mammals, such as some primates, the sacs descend seasonally to be retracted when the breeding season is over. They descend because the high body temperatures hinder the process of sperm maturation.

The length of time young mammals remain with the mother varies from group to group. Some are precocious, and able to fend for themselves when they are only a few days old, although they are usually cared for by the mother for several months at least. Others may not become independent for many years. This is particularly true of social mammals such as elephants which live in family groups and herds. The young stay with the mother for 12 to 14 years. Elephants and other community animals cooperate in finding food, avoiding danger and caring for the young. This life style necessitates some sort of code of behavior for herd members, which is learned from the parents and other members of the group.

Land-dwelling migrating mammals have been greatly reduced in number because of man's population of the globe. Those that still migrate include the caribou and reindeer, which inhabit the northern expanses of Canada and Arctic Europe.

One of the most significant features of mammals is the yolkless egg and the development of the fetus in the uterus. It is protected by the amniotic sac and the allantois, and is nourished by the placenta which carries oxygen and nutrients to it and removes waste.

Monotremes and marsupials

The classic definition of a mammal is an animal that gives birth to fully-formed young that are initially nourished on mother's milk. But there are two groups of mammals that defy this definition: the monotremes and the marsupials.

The monotremes (subclass Prototheria) lay eggs, and the marsupials (subclass Metatheria) give birth to embryonic young that continue their development within the protection of the mother's abdominal pouch. The six species of monotremes and 250 of marsupials together make up only about six per cent of all mammal species, the remainder of which constitute the placental mammals (subclass Eutheria).

Inhabiting the rivers of eastern Australia is the duck-billed platypus *(Ornithorhynchus anatinus).* This monotreme feeds off the river beds, probing for small crustaceans and worms with its sensitive bill. Unlike most mammals its warm-bloodedness is not fully developed. Its body temperature is lower than that of other mammals at 86°F (30°C), and is not constant, but fluctuates. It has an average life span of 11 to 12 years.

The koala *(Phascolarctos cinereus)* of eastern Australia feeds almost exclusively on eucalyptus leaves and the young bark of the trees. The female of this arboreal species usually produces a single offspring every alternate year, which she carries on her back for about a year. Koalas measure from 25 to 30 inches (64 to 76 centimeters) in length and can weigh 30 pounds (14 kilograms). On their front limb the first and second fingers are opposable with the other three, which aids them in grasping tree trunks and branches.

Monotremes

The monotremes are the platypus and the spiny anteaters. The most striking difference between them and other mammals is that they lay eggs covered by a leathery shell. This similarity to reptiles is counterbalanced by the mammalian one of feeding the hatched young on milk from mammary glands. These glands do not have teats, but emerge on the abdomen (platypus) or in a temporary pouch (spiny anteaters).

In monotremes, just as in reptiles, the ducts of the excretory system and the genital tract have a common opening known as the cloaca. It is this feature that gives monotremes their name, which means "one hole." The structures of the bones in the lower jaw and middle ear are similar to those of other mammals but the girdle of bones that supports the forelimbs is reptilian in form. In addition, the brain and circulatory systems of monotremes are mammalian but have some reptilian features. This peculiar mixture of characteristics may be an indication of an evolutionary link between the two groups.

There are two families of monotreme: Ornithorhynchidae (bird-noses), which contains a single species *(Ornithorhynchus anatinus),* the duck-billed platypus; and Tachyglossidae, the spiny anteaters, which consists of five species.

The platypus, which is found only in Australia, is a streamlined animal with a flat snout like a duck's bill. It lives in burrows which it digs in riverbanks or inherits from other animals. It hunts for food underwater, depending largely on the tactile sense organs on the soft edge of its bill to find its prey, which includes small animals such as larvae, earthworms and crustaceans.

The platypus has webbed feet, and dorsal nostrils on the tip of its bill through which it breathes while it floats on the surface of the water. When it dives, a protective skin fold closes over its ears and eyes and it can stay submerged for up to five minutes. Male platypuses have a venom spur on each hind leg, as do male spiny anteaters.

Spiny anteaters, or echidnas, occur in the sandy and rocky regions of Australia and New Guinea. They have a protective coat of short, sharp spines, rather like a hedgehog. These animals have no teeth but manage to feed on termites, ants and other insects by picking them up with their long, sticky tongue.

The female spiny anteater temporarily develops a pouch on her abdomen during the breeding season, and transfers her eggs into this pouch after laying. The young hatch about a week later, and feed from the milk ducts which open into the pouch. They stay in the pouch for six to eight weeks, until their spines have developed.

Marsupials

Monotremes are confined to six species but fossil records of the Eocene age, 55 million years ago, indicate that marsupials were once widely distributed and even more common than eutherians are. Now marsupials are restricted to North and South America (70 species) and Australia (more than 170 species). One reason for this reversal could be that marsupials have a smaller, simpler brain than the eutherians do and so stand a smaller chance of survival compared with the more intelligent eutherians.

The marsupials are an extremely diverse group and include arboreal, fruit-eating, grazing, burrowing, insectivorous and carnivorous types. There are six families of marsupial: opossums (Didelphidae), carnivorous marsupials (Dasyuridae), bandicoots (Paramelidae), rat opossums (Caenolestidae), phalangers (Phalangeridae), and kangeroos (Macropodidae).

The most distinguishing feature of a marsupial is the pouch, or marsupium, which contains the mammary glands, and which gives marsupials their name although there are some species in which it is not present. The embryo in some marsupials is supplied with nutrient from the wall of the uterus, whereas in others there is no placental connection between the uterus and the embryo. In all species, the embryo remains in the uterus for a very short period and develops quickly.

At birth, the embryo is tiny — 1.2 inches (3 centimeters) long at most, and many are no longer than a grain of rice. Called a neonate, its forelimbs and nervous center are well developed, but the hindlimbs are mere buds. It crawls, unaided, from its mother's reproductive tract to her pouch where it latches on to a teat which injects milk into the young animal's mouth. A baby kangaroo or joey, for example, remains in its mother's pouch for four or five months, and will return to feed or seek refuge until it is a year old.

The kangaroos, perhaps the best known marsupials, have forward-opening pouches but the form of pouch varies between species. For example, wombats and phalangers, including the koala *(Phascolarctos cinereus)*, have pouches which open to the rear, an advantage for the young of digging animals although it means the mother cannot clean the pouch.

The marsupial reproductive system differs from that of eutherians in various other ways: for example, the females have a double uterus and double vagina, and in many species the males have a forked penis, with the testes in front of it. But from the similarities in their reproductive systems it seems that the marsupials and the monotremes form an intermediate group between the reptiles — which lay eggs covered with shell and do not have mammary glands — and the eutherians — whose young remain in the uterus until they have developed to an advanced stage, when they are born and then fed with milk from the mammary glands.

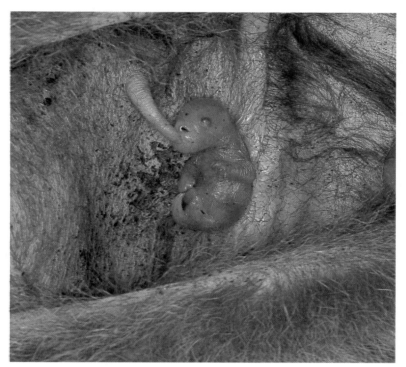

The young of marsupials are known as neonates. In kangaroos (Macropodidae), a single neonate emerges from the mother's cloaca about 33 days after conception. Its forelimbs are well developed, but its hindlimbs are mere buds. It struggles through the mother's fur to her pouch where it latches on to one of the teats. About six months later the joey forays from the pouch.

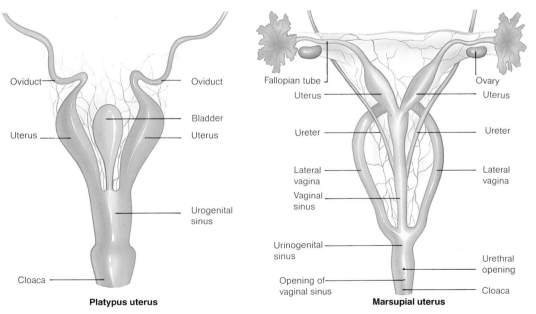

Platypus uterus

Oviduct — — Oviduct
— Bladder
Uterus — — Uterus
— Urogenital sinus
Cloaca —

Marsupial uterus

Fallopian tube — — Ovary
Uterus — — Uterus
Ureter — — Ureter
Lateral vagina — — Lateral vagina
Vaginal sinus
Urinogenital sinus — — Urethral opening
Opening of vaginal sinus — — Cloaca

The uteri of monotremes and marsupials are unlike those of eutherians in that they have two uteri and one urinogenital opening. In both the eggs are fertilized in the oviducts. In the platypus they are covered with a leathery shell and two eggs are usually produced at a time. In the marsupials, the sperm go up two vaginal canals to fertilize the eggs; the neonates bypass the canals and come down the vaginal sinus. Opossums carry several embryos in both uteri simultaneously, but kangaroos carry only one.

Insectivores

The Insectivora is probably the most ancient order of placental mammals; these insect-eating mammals are known to have lived as far back as 100 million years ago, sharing their world with the dinosaurs. They all share certain basic features, but each species has specialized for a different way of life so that they are superficially dissimilar. With bats (order Chiroptera), which are generally considered to be descended from Insectivora stock, flying lemurs (Dermoptera), anteaters, sloths and armadillos (Edentata), pangolins (Pholidota) and some primates, they form the bulk of a large group of mainly insect-eating mammals.

Insectivore classification

Today, Insectivora is the third largest order of mammals, comprising four suborders and about 400 species. The suborder Lipotyphla contains most of the insectivores, and includes mouselike creatures such as solenodons, tenrecs, moles, desmans and golden moles; hedgehogs, moonrats and shrews. The related water- or otter-shrews have been placed in their own suborder — Zalambdodontia — because their teeth differ from the rest of the insectivores. The elephant-shrews have also been separated, into the suborder Macroscelidea, for the reason that they hop on their back legs. The suborder Dermoptera contains the flying lemur — the most accomplished of gliding mammals.

Insectivores occur throughout the world except for Australia, New Zealand and some oceanic islands. Typically they are small, nocturnal mammals with elongated narrow snouts. They are distinguished from other small mammals by the fact that they have five digits on each limb — rodents, for example, have four or less. Each digit has a distinct claw. The body is covered with short, dense fur; in some species, such as hedgehogs and tenrecs, some of the hairs are developed into spines. They have small ears and small eyes with poor vision; their senses of smell and hearing, however, are usually sharp. The brain has a primitive structure and the placenta is simpler than that found in most other mammals. Insectivores have up to 44 teeth — the largest number normally found in placental mammals. These usually have sharp cusps which enable them to slice their prey. Insectivores include

ground-living, burrowing, climbing and even semiaquatic species. They feed on insects, grubs and snails and, occasionally, on helpless vertebrates such as the young of ground-nesting birds. Many insectivores are extremely active and therefore need to be refueled constantly by large quantities of food — sometimes equaling the animal's total body weight each day.

Habitats and life styles

Of the two species of solenodon, one was last seen in Cuba in 1909 and is now thought to be extinct. The other lives on Haiti — *Solenodon paradoxus.* This animal is about the size of a rat and looks like a shrew, with its pointed, almost naked, snout. It uses this sensitive snout to root about on the ground for invertebrates and plant material; it also eats lizards, frogs and small birds — like some other insectivores, it produces a toxic saliva from a gland in the lower jaw.

The tenrec family (Tenrecidae) includes about 20 species, all found on Madagascar and the Comoro islands. They are primitive animals and retain some reptilian features, such as the cloaca — a common opening for urogenital and anal systems — but are most interesting in that they have adapted to fill a wide range of habitats in Madagascar. Some resemble shrews; those of the genus *Setifer* resemble hedgehogs and have sharp spines; others are similar to mice and moles. Their main diet consists of small invertebrates and plant material.

The diversity of insectivores is reflected in their limb structures. All have basically pentadactyl (five-toed) limbs, modified to suit their life styles. Shrews walk on their toes, whereas a mole's digits form a broad spade for digging. Anteaters have an enlarged third digit with a curved claw for breaking open ant and termite nests. Bats have four elongated digits on the forelimbs, which support the membranous wing.

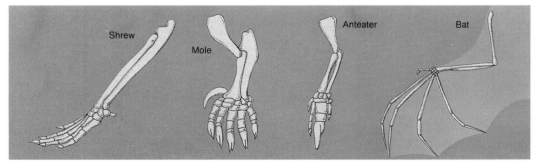

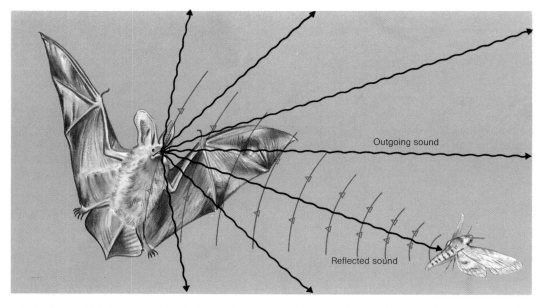

Most bats (far left) eat insects and, like this noctule *(Nyctalus noctula)*, hunt their prey at night using echo-location (left). While flying through the air at up to 18 miles (24 kilometers) per hour they generate high-frequency sounds and detect prey or obstacles by the echoes they reflect. The range of detection is, however, limited to 3 feet (1meter) or so but is sufficiently discriminating to enable some bats to catch fish by detecting the ripples they make on the water surface.

The family Talpidae consists of about 29 species of mole, shrew-mole and desman, which occur widely in the Northern Hemisphere. Most are burrowers and are highly modified for an underground existence with powerful spadelike front feet, minute eyes with poor vision hidden in the fur, and no external ears. The short fine hairs on the body can be brushed in any direction, enabling them to reverse easily in their tunnels. The desmans,which are the largest of this group, lead a semiaquatic life in, or near, ponds and streams, and feed on aquatic invertebrates and fish. They use their long noses as snorkels, turning them up so that they stick out of the water.

Golden moles, which resemble true moles, form the family Chrysochloridae, and are found in a range of habitats in Africa, from desert to forest. They have a blunt snout covered with a leathery pad, the eyes are vestigial and the front feet are armored with powerful claws which are used for efficient burrowing. Golden moles are diurnal and feed on a range of underground invertebrates. Some desert-dwelling, sand-burrowing types, such as the desert golden mole *(Eremitalpa granti),* even capture and kill burrowing reptiles.

The hedgehog family (Erinaceidae) includes the familiar hedgehogs, with their covering of sharp, tough spines, and the more rodent-like, long snouted moonrats *(Echinosorex gymnurus).* There are 17 species in all, found in Europe, Asia and Africa. They are nocturnal, feeding on a range of invertebrates, but also on carrion and small mammals.

The shrews of the family Soricidae are the most successful insectivore group in terms of numbers; there are about 250 species living almost everywhere in the world. Most are small — the pigmy shrew *(Microsorex hoyi)* is the smallest and weighs less than 0.08 ounce (2.5 grams). Seldom seen, shrews forage on the ground, moving under logs, plant debris and into crevices in search of invertebrate prey. Some are able to subdue larger animals such as mice and frogs because, like solenodons, they also have a highly toxic substance in their saliva. The prey is rendered helpless by just a few bites from these venomous animals. Shrews communicate by high-pitched squeaks and by noises of too high a frequency for the human ear to hear.

The water- or otter-shrews of the order Zalambdodontia are different from other insectivores in that they have four-cusped molars, the cusps being arranged in a W-shape. This formation is known as zalambdodont dentition and is almost identical with that of the earliest placental mammals. They lead a semiaquatic life in streams, rivers and swamps, feeding on fish, crabs and frogs. The giant otter-shrew *(Potamogale velox)* is the largest living insectivore, with a length of about 24 inches (60 centimeters). With its sleek dense coat and powerful tail it resembles a small otter.

Hedgehogs are mainly nocturnal animals and predominantly insectivorous, although they have been known to feed on carrion and small mammals. When food is scarce in winter, they hibernate.

The elephant-shrews differ from other insectivores in that their hindlimbs are adapted for hopping. There are about 18 species, found in Africa, in habitats ranging from semidesert to forest. They have large eyes and ears and are further distinguished by their extremely long, narrow snout. It is the supposed resemblance of this snout to an elephant's trunk that led to the origin of its common name. These animals forage on the ground using their mobile snouts and long-clawed front feet to search for small invertebrates.

Flying lemurs

The two surviving species of flying lemur (also known as colugos) represent an early specialized development of the basic insectivore type. These southeast Asian tree-dwelling mammals are not really able to fly, but glide among the treetops by means of membranes at the sides of the body. These membranes extend from the neck to the fore feet, the hind feet and the tip of the tail. Normally held folded in at the sides, the membranes become parachutelike when the animal extends them by stretching out its limbs. Flying lemurs are excellent climbers and spend the day hanging by their clawed feet from a branch. At night they glide from tree to tree feeding on leaves, buds, flowers and fruit. Their jaw is unusual among mammals in that its lower incisors have become "comb teeth." Used for grooming the fur, the teeth are similar to those of the lemurs — an example of convergent evolution from two separate family groups.

Bats

Descended from the insectivores, but far more successful, the bats are second only to rodents in terms of numbers. At least 950 species are known from almost all parts of the world. The order is divided into two suborders: Megachiroptera contains one family, the Old World fruit bats, which feed primarily on fruit; Microchiroptera contains the other 17 families, most of which are largely insect-eating.

The main reason for the enormous success and diversity of bats is their use of an otherwise unoccupied ecological niche — that of nighttime flying mammals. No other mammal can actively fly and no other creatures, even birds, can surpass the bats' mastery of the night skies, achieved by their use of a sonar sensory system for navigation and prey-finding.

To enable them to fly, bats have evolved extensive wings that consist of thin skin flaps stretched between the enormously elongated fingers of the hands and between the forearm and the much smaller hind legs. In most species, the skin flap also extends backward from the legs to the tail.

The three-toed sloth *(Bradypus tridactylus)* is found in the Amazonian forests where it feeds on leaves and rarely descends to the ground. Its three digits are bound for almost their entire length by skin and have long, curved claws.

The tamandua, or collared anteater *(Tamandua tetradactyla)*, is a mainly nocturnal, arboreal species found in South America. Its body length reaches 23 inches (58 centimeters). This animal walks on the outside of its hands to prevent the tips of its curved claws from digging into its palms.

The skin is an extension of the back and belly. In flight bats are exceptionally maneuverable and when they alight they hang upside-down from perches with the aid of clawed hind feet.

Skill in flight and echo-location has enabled Microchiropteran bats to adapt to a wide range of feeding strategies apart from the most important one of capturing night-flying insects. Some feed on the nectar and pollen of night-flowering plants and are important pollinators of those plants. Others feed on small invertebrates and mammals, and even capture fish in swoops to the water surface. The vampire bats (Desmodontidae) have almost chisellike incisors which cut a narrow groove into the skin of sleeping mammals and birds, and sever capillaries, which bleed freely. They lap up the blood that flows from these wounds.

Armadillos, anteaters and sloths
The order Edentata today contains only 29 species in three families: armadillos (Dasypodidae), anteaters (Myrmecophagidae) and sloths (Bradypodidae). All have a reduced number of teeth, or none at all.

The 20 species of armadillo of the New World are nocturnal and feed primarily on insects as well as other small invertebrates and vertebrates and, in some cases, carrion. Their bodies are protected by an armor of horny plates on the skin, leaving only the limbs and underside vulnerable to attackers. Some species of armadillo curl up into a ball to protect these parts of the body. They have simple, cylindrical teeth which are gradually shed with age. Their limbs are sturdy and powerful, with large strong claws.

Also living in the New World — in Central and South America — are four species of anteater, all of which are highly specialized for feeding on ants and termites. The anteater has a long snout and no teeth, but a long sticky tongue which is flicked out to capture insect prey. The giant anteater *(Myrmecophaga tridactyla)* is ground-dwelling, but the other species spend at least some of their time in the trees.

The five species of sloth are so adapted to life in the trees that they can barely move on land. These creatures have earned their name from their slow movements. For this reason they rarely descend to the ground because they would make easy prey. They live in forests in Central and South America and spend much of their time hanging upside-down from their large hooklike claws. There are two genera — Bradypus, the three-toed group,

and Choloepus, which are two-toed. The three-toed species tend to be slower than the two-toed and their grooved hairs carry algae which give them a green color.

Pangolins
The pangolins (order Pholidota), or scaly anteaters, of Asia and Africa lead a similar life to the American anteaters. They have no teeth and feed largely on termites, ants and other insects which they capture with deft movements of their long tongue. The largest of them can protrude its tongue 16 inches (40 centimeters) out of its mouth. Like anteaters, pangolins have a gizzard-like stomach which grinds down the hard exoskeletons of the insects they eat.

Scaly anteaters, or pangolins *(Manis* spp.), are distinguished by their scaly skin, which resembles the bracts of a pine cone. These arboreal mammals have a prehensile tail and a flexible body which can roll into a ball. Like the anteaters, they have no teeth, but have strong claws for tearing open termite nests.

Fact entries

Echo-location is a means of establishing the position of objects by monitoring the sound waves that bounce back off them, and is used by bats. These animals emit ultrasound — high energy pulses or whistles of sound at frequencies of up to 150KHz, which are far above the upper frequency limit of the human ear, which is around 20KHz. Bats produce the sounds with their large, specialized larynx and project them either through the open mouth or through the nose, which is designed as a transmitter, with leaf-like appendages. They emit these sounds, while they are in flight, in a rapid series of squeaks, each lasting for only about $\frac{1}{500}$ of a second, at a rate of about 50 per second. Ultrasound bounces off objects in a similar manner to light. Bats receive these reflected noises with their large, highly modified ears like dish radio telescopes, which receive in a similar way. The returning sound signals are analyzed by the bat's inner ears and enable it to estimate with extraordinary accuracy the range, direction, nature and speed of objects flying through the air. Bats are thus able to avoid danger and, just as important, locate food.

Primates

Man and his closest relatives in the animal kingdom belong to the order Primates. The name, which means "first ones," is apt because early members of this group were contemporaries of the dinosaurs, although they were not the earliest placental mammals. The primates began to diverge from the other mammalian orders about 80 million years ago, then differing from them very little in appearance and behavior, but bearing the evolutionary potential whose expression since that time has produced the diverse forms of their descendants today.

The present order is divided into two suborders: the Prosimii, including tree-shrews, tarsiers, lemurs, indris, aye-ayes and lorises; and the Anthropoidea, consisting of the New World monkeys (tamarins and marmosets, and cebid monkeys), the Old World monkeys (including the macaques, baboons and colobus monkeys), the lesser apes (gibbons and siamangs), and the great apes (orangutans, chimpanzees and gorillas). Among the smallest primates is the pygmy marmoset *(Cebuella* sp.), whose body length measures about 5.5 inches (14 centimeters) the largest is the gorilla *(Gorilla gorilla)*, which stands at about 5.8 feet (1.75 meters) high and weighs about 600 pounds (275 kilograms).

Distribution

The prosimian group is dominated by the lemurs and their relatives, which are concentrated in Madagascar and the nearby Comoro Islands. This is because about 50 million years ago, the earliest lemurlike prosimians on Madagascar were separated from Africa by the Mozambique Channel. There were few predators on Madagascar and so they were able to diversify. Those that remained on the mainland were forced to live a nocturnal life by the more versatile Old World monkeys and apes that developed. These nocturnal prosimians include bushbabies *(Galago* spp.) in tropical Africa, and lorises (Lorisidae) and tarsiers (Tarsidae) in tropical Asia. Prosimians failed to reach Australia because it was isolated in the Pacific well before their evolution. Tarsier and lemurs once existed in North America, but were ousted by the New World monkeys that evolved there. The success of the primates meant that some species increased in size and moved from the ancestral forest habitats. Today the largest primates spend little time in the trees. Primates, whatever their way of life, have remained essentially animals

Lemurs, such as the ring-tailed species *(Lemur catta)*, inhabit Madagascar. They were revered by the people of the islands as the home of the spirits of the dead, but since the imposition of Christianity by European colonists they have been killed for food.

The Primates consist of two suborders: the Prosimians, which include tree-shrews and lemurs, and the Anthropoids, which contain New World monkeys, Old World monkeys and hominoids.

PROSIMIANS	NEW WORLD MONKEYS	OLD WORLD MONKEYS	HOMINOIDS
Tree-shrew (*Tupaia* sp.)	Marmoset (*Callithrix* sp.)	Mandrill (*Mandrillus* sp.)	Gibbon (*Hylobates* sp.)
Lemur (*Lemur* sp.)	Spider monkey (*Ateles geoffroyi*)	Colobus monkey (*Colobus* sp.)	Gorilla (*Gorilla* sp.)

of the tropics, rarely being found where warmth and food are not constantly available.

Evolutionary trends

The earliest primates are thought to have fled from their large neighbors by scrambling into the trees. This arboreal tendency and its adaptations has remained with the primates throughout their evolutionary history.

The more distinctive developments among the primates include the evolution of hands and feet equipped with grasping fingers and toes. This involved the separation of the digits and the development of musculature which enabled them to fold round a branch or other objects. The first digit (thumb or big toe) became widely separated from the others, but could, in many cases, be folded across the palm of the hand, providing a manipulative grasp. The tips of the digits became flattened, and nails — rather than claws — developed on their dorsal surfaces. Friction pads of deeply folded skin developed to maintain a hold on smooth or slippery surfaces.

Another important development was the reduction in the size of the snout, which allowed greater room for the eyes to move forward and evolve stereoscopic vision. This was particularly important for judging distances when leaping among the trees.

High activity levels and manipulative skills require enlarged brain centers and the primates tended to develop big brains in comparison with their body bulk. In general, this leads to a higher level of intelligence, but is not an absolutely rigid ratio.

In addition, the development of the embryo in the womb becomes more complex in primate evolution, and takes longer. This correlates with a prolongation of parental care. Most primates are highly social animals, leading a life of complex

interactions between members of a group. With social safety and prolonged care, education of the young can develop. Strong links are forged between an infant and its mother in the early stages of life. Later, bonds are made with playmates and other members of the group.

Each of these (and other) trends has progressed to different points in different primates but is generally less advanced in the prosimians than in the higher primates, the anthropoids.

Prosimians

The inclusion of the predominantly ground-dwelling tree-shrews (Tupaiidae) in the primate order is debatable because they have few primate characteristics. Even so, these animals, which inhabit the forests of tropical eastern and south-eastern Asia, are thought by some zoologists to be the earliest relative of primates. The Latin name Tupaia is derived from the Malay word for a squirrel, which tree-shrews resemble in size and, to some extent, in appearance. They are small, often bushy-tailed, clawed creatures, but they differ greatly from the rodents in their dentition, which

The slow loris (*Nycticebus coucang*) is, as its name implies, a slow-motion climber. The opposable thumbs and big toes of this animal give it a powerful grip as it pulls itself along, hand over hand, suspended under branches. It catches by stealth, and eats birds and other small animals.

The prototype primates began to separate from the other mammalian orders about 80 million years ago. Today the order contains about 170 species — 54 among the prosimians and 117 in the anthropoids.

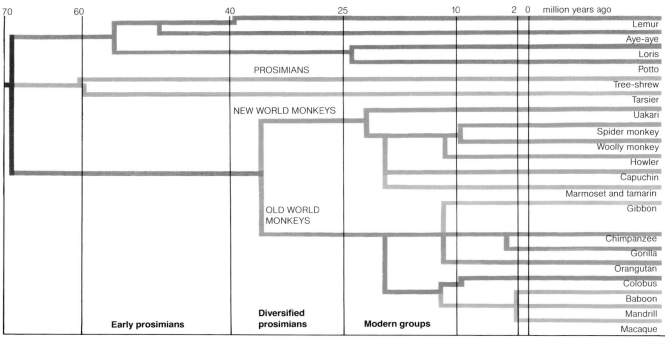

| 70 | 60 | 40 | 25 | 10 | 2 0 | million years ago |

- Lemur
- Aye-aye
- Loris
- Potto

PROSIMIANS

- Tree-shrew
- Tarsier

NEW WORLD MONKEYS

- Uakari
- Spider monkey
- Woolly monkey
- Howler
- Capuchin
- Marmoset and tamarin
- Gibbon

OLD WORLD MONKEYS

- Chimpanzee
- Gorilla
- Orangutan
- Colobus
- Baboon
- Mandrill
- Macaque

Early prosimians — **Diversified prosimians** — **Modern groups**

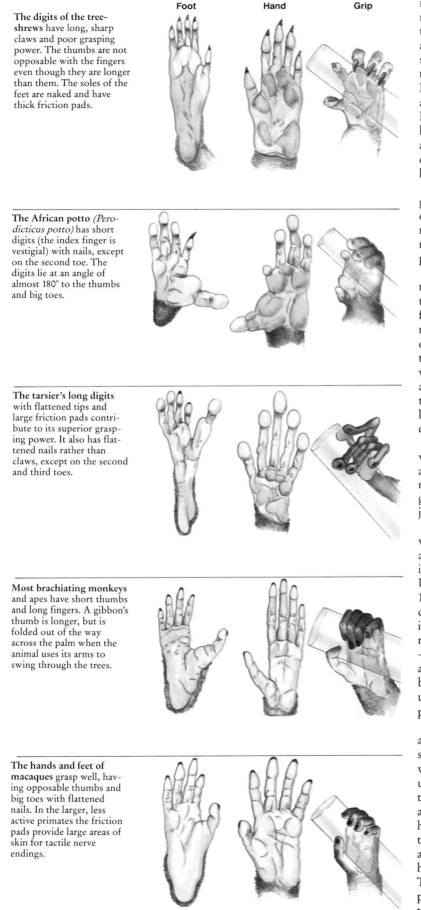

Foot   Hand   Grip

The digits of the tree-shrews have long, sharp claws and poor grasping power. The thumbs are not opposable with the fingers even though they are longer than them. The soles of the feet are naked and have thick friction pads.

The African potto *(Perodicticus potto)* has short digits (the index finger is vestigial) with nails, except on the second toe. The digits lie at an angle of almost 180° to the thumbs and big toes.

The tarsier's long digits with flattened tips and large friction pads contribute to its superior grasping power. It also has flattened nails rather than claws, except on the second and third toes.

Most brachiating monkeys and apes have short thumbs and long fingers. A gibbon's thumb is longer, but is folded out of the way across the palm when the animal uses its arms to swing through the trees.

The hands and feet of macaques grasp well, having opposable thumbs and big toes with flattened nails. In the larger, less active primates the friction pads provide large areas of skin for tactile nerve endings.

is adapted to a diet of insects, fruit and small birds, rather than the specialized gnawing and chewing teeth possessed by squirrels. Unlike squirrels they are nocturnal and have highly light- but not color-sensitive eyes. These are situated on each side of the head so that there is no stereoscopic vision. But like squirrels, they have prominent snouts and depend heavily on their good sense of smell. In their favor as primates, however, tree-shrews have moderately large eyes, the sockets of which are encased in bone; they have bony ridges (post-orbital bars) surrounding the eye sockets, and have primatelike ears.

Tree-shrews are among the few primates that generally produce twins. Each of the twins develops in one of the two long branches of the bicornuate uterus common to the primitive placental mammals. Gestation lasts about six weeks and parental care is minimal.

The status of the tarsiers *(Tarsius* spp.) as primates is certainly less controversial than that of the tree-shrew. These arboreal animals inhabit the forests of southeastern Asia and one of their most notable features is the greatly flattened tips of their digits. These allow the tarsiers to be able to jump and support themselves even on smooth vertical surfaces. The ability to grasp and climb among smooth branches, which is where most trees produce their most succulent buds and leaves, enabled the early primates to colonize forest habitats unused by other mammals.

Travel between high fruit-bearing branches was made possible by the development of leaping as a means of progression. The tarsier gets its name from its means of leaping — its greatly elongated anklebone or tarsus which allows it to jump at least 6 feet (2 meters).

The tarsier has a relatively flat face and forward-facing eyes with stereoscopic vision, which aids it when jumping. Its enormous, fixed, bulging eyes are the tarsier's most remarkable feature. Unable to move them, it can swivel its head a full 180° to the left or right. The size of its eyes may contribute to its good night vision, but a more important factor is the abundance in the retina of rods — which operate well at low light intensities — and absence of color-sensitive cones. This arrangement is typical in nocturnal vertebrates because color vision and fine resolving are less useful in the dark. Its eyes lie in eye sockets completely encased by bony ridges.

The largest number of prosimians are found among the lemurs. These animals, which are superficially monkey-like, are arboreal, have forward-facing eyes with less well-developed binocular vision than other primates, but with a tapetum (a reflector at the back of the retina) which allows them to see at night. Like the tarsier, they have nails rather than claws on their digits (with the exception of the clawed second toe), which allow them to grip more easily. Unlike the tarsier, however, they have pseudo-opposable thumbs. Their nose is fox-like in shape, although not as pointed as the tupaia's. Scent — an important feature in the life of these animals — is secreted from

scent glands to convey a complex language of signals — in aggression and marking territory, for example. The lower incisor teeth are modified to form a comb which is used for cleaning the fur (as is the claw on the second toe).

Related to the true lemurs and also found in Madagascar are the indris and the aye-aye *(Daubentonia madagascariensis)*. The aye-aye was once thought to be a rodent, mainly because its curved, chisel-like incisors continue to grow throughout its life. But if they did not, the teeth would be worn away because the aye-aye tears at rotten wood and gnaws at tree branches for hidden grubs detected by their sensitive ears, before it dislodges the prey with its extraordinarily long, thin, hooked middle finger. The finger is twice as long as the others, and has a sharp claw.

In Africa the prosimians are most represented by the bushbabies. These animals are fast-moving with tremendous jumping power, and are nocturnally active on the ground or in the shrub layer of the forest. Their eyes are large and forward-pointing and the snout is reduced, giving them good binocular vision. They have the curious habit of urinating on their hands and feet, probably for territorial claim, spreading their scent as they leap about. Bushbabies are unlike the lorises and pottos which, although similar in appearance, travel differently. They exert a vice-like grip on a branch with their hands and feet while they move slowly upside down along the branch.

New World monkeys
Most of the primate group inhabiting the New World are considered to be more primitive than the other anthropoids. They include two families — the Callitrichidae, which consists of the 21 species of marmosets and tamarins, and the Cebidae, which contains 26 species, including howlers

(Alouatta spp.) and capuchins *(Cebus* spp.). They are confined to tropical forest areas and are, in general, diurnal animals. Only the douroucouli, or owl monkey *(Aotus trivirgatus)*, is nocturnal.

New World monkeys are termed platyrrhine, because they have flat noses, with widely spaced nostrils. They have better stereoscopic vision than the prosimians and also have relatively sophisticated grasping faculties. These two complementary developments may explain, in part, why they displaced the prosimians.

The callitrichids are the smallest of the New World monkeys, measuring usually not more than 10 inches (25 centimeters). They have curved, claw-like, keeled nails on all their digits except for the big toe (which has a flattened nail). These claws give them sufficient grip as they travel along branches on all fours. Unlike the cebid monkeys, they have non-opposable thumbs and grasp objects between the fingers and the palm of the hand. Like the tree-shrews the female marmosets usually have twins. They are carried and cared for by their father, who hands them over to his mate to be suckled.

All the cebid monkeys have better manipulation than the callitrichids — the capuchins being the most skilful. Larger and heavier than the marmosets and tamarins, they have to hold on to branches rather than merely balance on them. Like the callitrichids, some cebid monkeys live in family groups; others, however, are found in large, multi-male groups. In a few species the males carry and care for the young, but they are generally tended entirely by the females.

The prehensile tail
In addition to their manual dexterity some of the South American monkeys are further supported by a prehensile tail. This tail is possessed only by the howlers, the spider monkeys *(Ateles* spp.),

The tarsiers (Tarsiidae, left) are small, nocturnal animals, which can rotate their heads by 180° in order to see because their eyes are fixed. This curious feature led local Borneo inhabitants to believe that if they saw a tarsier while warfaring, they would be successful in taking many enemy heads.

The tiny marmosets of the New World often have large tufts of fur which conceal their ears, such as this golden lion tamarin *(Leontideus rosalia)*. Unlike the rest of the New World monkeys, marmosets and tamarins have long, keeled claws on their digits, and their thumbs are not opposable.

woolly monkeys *(Lagothrix* spp.) and the woolly spider monkey *(Brachyteles arachnoides).* In howler and woolly monkeys the tail is used as an extra hand, although it reaches its fullest development in the spider monkey, which can support its whole weight from the tail. The end of the underside of the tail is naked and has ridged skin, which provides a better grip. It is also very sensitive and can pick up even tiny objects.

Anatomy and the senses

The spider monkeys are typical of New World monkeys in that they move through the trees by jumping from branch to branch as well as swinging from their arms. This latter method of loco-

motion is called brachiation, the supreme exponents of which are the Old World apes, the gibbons *(Hylobates* spp.). Like many brachiators, spider monkeys have hook-like hands with reduced thumbs, and arms which are much longer than the legs. Senses such as the olfactory one are important to these monkeys. Marmosets and tamarins mark their territory and points of reference with scent released from glands in the scrotal and anal regions; and the capuchins rub their chest glands on branches for the same purpose and also urinate into a cupped hand with which they anoint a foot before moving along their territory.

Color vision and visual acuity are also significant to the New World monkeys (apart from the nocturnal douroucouli). The bald uakaris *(Cacajao* spp.), for example, have brilliant red faces set off by their white or red-brown fur. As well as species recognition, color also plays a role in food recognition' — those with color vision can estimate by looking at it the ripeness of fruit or the freshness of foliage.

The brains of the New World monkeys are much larger and more complex than those of the prosimians. The greater weight of the brain is largely attributable to the expansion of the cortex, which coordinates sensory and motor functions and controls memory and intelligence. This overall trend in the dominance of the cortex (corticalization) is more advanced in the cebid than in the callitrichid monkeys.

Males are, in most cases, larger than the females. Like most other animals they are territorial — the male howlers are most notable for the method in which they proclaim their dominance over their territory. They have an enormous hyoid bone which makes a cup-shaped sounding box in the throat. This air sac is inflated to resonate sounds that carry a distance of about 2 miles (3.2 kilometers).

Few New World monkeys show evidence of

Prehensile tails are particular to only a few of the New World monkeys, such as this howler *(Alouatta* spp.). These are the largest of the American monkeys, having a body length of about 24 inches (61 centimeters) and weighing about 18 pounds (8 kilograms). They therefore can use their tails only as an extra hand, whereas the spider monkeys, which are slighter, can hang their whole body from the tail.

Coloration is important in displays among many monkeys. The red-faced uakari *(Cacajao rubicundus)* is most notable for its bright-red face and almost bald head. The uakaris are the only New World monkeys with short tails. They inhabit the upper layer of the Amazonian forest in troops and rarely descend to the forest floor.

menstrual cycles; actual menstruation is rare and minimal, and swelling of the female external genitals is minor. Their uteri have no uterine horns, unlike the prosimians. Gestation lasts about 20 to 25 weeks, and their infantile, juvenile and adult phases are longer than those of the prosimians. The extension of the postnatal phases is a clear trend further developed in the higher primates and is critical to learning and social behavior.

Old World monkeys

The monkeys of the Old World are found in Africa and the warmer parts of Asia and, with human help, the Barbary apes survive in Gibraltar. Old World monkeys live across a wide range of habitats and are far more varied in their ways of life than are the cebids or marmosets. Some inhabit semiarid areas beyond forest boundaries where they are forced to hunt on the ground for any food they can find.

These monkeys are all grouped in one family — the Cercopithecidae. This family is divided into two subfamilies — the Cercopithecinae, which includes the macaques *(Macaca* spp.), the African baboons *(Papio* spp.), and the guenons *(Cercopithecus* spp.); and the Colobinae, which includes the langurs *(Presbytis* spp.), the Colobus monkeys *(Colobus* spp.) and the proboscis monkey *(Nasalis larvatus)*. The subfamily divisions reflect different eating habits. Most colobine monkeys are vegetarian, whereas the cercopithecine monkeys, most of which are ground-dwelling, are generally more omnivorous and greater opportunists, eating what they find. They have cheek pouches in which they store food that they do not eat immediately.

The most obvious difference between Old and New World monkeys is that the nostrils of the former are closer together and point downward rather than sideways. They are therefore known as narrow-nosed, or catarrhine, monkeys. Another distinguishing feature of all catarrhine monkeys is the presence of bare patches of hardened skin on the rump, called ischial callosities. These patches have neither nerve nor blood supply and allow the monkeys to remain in a sitting position for long periods without serious discomfort — while sleeping, for example.

Hands, limbs and locomotion

Many Old World monkeys have fine manipulative skills which are associated with true opposability

of the thumb — its tip can be pressed against at least one finger, or its tip, on the same hand. True opposability depends on the movement of the carpo-metacarpal joint and development of the thenar muscles at the base of the thumb, rather than on the movement of the next joint up in the thumb, the metacarpo-phalangeal (which confers pseudo-opposability, as in New World monkeys). The predominantly ground-dwelling baboons and mandrills *(Mandrillus* spp.) have relatively long, highly opposable thumbs and use them to good effect in pulling up grass and other plants. Colobus monkeys, in contrast, grip between their fingers and the palms of their hands because, like the spider monkeys of South America, their

Tree-shrew

Lemur

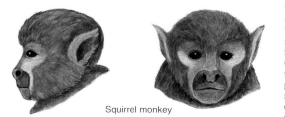

Squirrel monkey

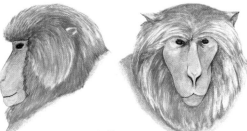

Macaque

Gorilla

One of the first adaptations of the early primates to an arboreal life was the reduction of the long snout found in the tree-shrews (tupaia) which, despite their name, are mostly ground-dwelling. This reduction enabled the eyes (which are on the sides of the tupaia's head) to move closer together and achieve binocular vision — essential for judging distances accurately when leaping. The lemurs are more arboreal than the tupaias are, but still spend time on the ground. Their muzzle is reduced to a fox-like snout and their eyes are closer than are the tupaia's, but their vision is not as well-developed as that of the squirrel monkey *(Saimiri sciureus)*. As with all New World monkeys, these are platyrrhine because their faces are flattened, their nostrils are set wide apart and face sideways. In contrast, the Old World monkeys, such as the macaques, are catarrhine — their nostrils are close together and point downward. The great apes, such as the gorilla, have developed heavy, jutting jaws and a large brain case.

The limb lengths of primates vary. The woolly spider monkey and the gibbons brachiate and their arms are longer than their legs, although more so in the gibbons, which rely solely on their arms whereas the spider monkeys also use their prehensile tail. Gorillas' arms are slightly longer than their legs.

menon called sexual dimorphism. An average female, for example, is only three-quarters of the weight of a male. In the patas and the olive baboon *(Papio anubis)*, the female is only about half the weight of the male.

Male physical prowess is important both inside and outside the group — in determining the social pecking order and in defense of the troop. Many of the open-country monkeys are at risk from predators and their social organization helps them to minimize the danger.

The powerfully built male baboons, geladas and mandrills possess fearsome canine teeth, whose display rather than actual use is often enough to discipline troop members or to frighten off predators or strangers. Their teeth are housed in prognathous jaws but the monkeys still have good binocular vision, because the muzzle forms a shelf over which the eyes can focus well.

The visual cortex is necessarily well-developed in higher primates because their recognition of and communication by color is extremely important to them. Mandrills, among the Old World monkeys, have brilliant blue and red facial and genital markings.

thumbs are very short, or absent. This feature is the derivation of their name, which means "maimed." Also like the spider monkeys, they brachiate through the forest.

The wooded savannas and open dry grasslands of subsaharan Africa are the home of primates which tend to move on all fours. The patas monkeys *(Erythrocebus patas)* have exceptionally long limbs relative to their trunk, enabling them to sprint through the grass with long, bounding strides, faster than any other primate.

Anatomy and behavior

The males and females of cercopithecine monkeys are very different in appearance — a pheno-

Reproduction

Female reproductive behavior is much the same in Old and New World monkeys, in being seasonal. Many cercopithecoid females, however — baboons, mandrills, some of the macaques and colobus monkeys — clearly advertise their peak of fertility to the males by pink, swollen genitalia. They also menstruate more heavily than the American primates and have regular menstrual cycles. Gestation and the postnatal phases are longer in the Old than in the New World monkeys.

The apes

The last group of the primates, the superfamily Hominoidea, includes the great apes. The

Macaques are the most widespread of monkeys. The Japanese macaques *(Macaca fuscata)* are some of the few Old World monkeys found in areas where heavy frosts and snow are frequent. They have adapted by growing long beards and thick fur. These animals show remarkable ingenuity and exercise of choice in their eating habits. This mother and her infant are washing sweet potatoes in seawater because they prefer salted potatoes. They are also known to throw rice grains into the water to wash away the husks.

main characteristic that distinguishes apes from monkeys in their lack of an external tail (although as with man, there are small internal tail bones). Like the brachiating monkeys those apes that brachiate have very long arms, which exceed the length of their legs.

The gibbons and the siamangs *(Symphalangus syndactylus)* are often referred to as the lesser apes because they are smaller than the others. They are both Asian species; the gibbons populate the tropical forests of southeastern Asia, Malaysia and Indonesia, whereas the siamang is confined to those of Sumatra and mainland Malaysia. The other Asian ape is the orangutan *(Pongo pygmaeus)* of Borneo and Sumatra; together with the gorilla *(Gorilla gorilla)* and the chimpanzee *(Pan* spp.), of the African tropical forests, they constitute the great apes.

Locomotion and anatomy
The Asian apes are the most accomplished brachiators of all the primates. The light siamang, which may weigh up to about 28 pounds (13 kilograms), and the still lighter gibbons are tremendously agile in the trees and rely almost exclusively on true brachiation to move about. Effective brachiation requires the smooth rhythmical transfer of the body weight from one arm to the other, as the hands alternately curl around and release a branch. But this method of locomotion is not without its dangers because many animals fall at some time and a high proportion of skeletons studied show evidence of broken bones. The brachiating apes have retained their thumbs, unlike the spider and colobus monkeys, even though these are short in relation to their fingers and the rest of the hand. The hand assumes the shape and function of a hook during brachiation, when the thumb plays no role at all, but is folded across the palm. On the ground gibbons and siamangs can run upright holding their long arms out of the way and to balance themselves.

In contrast to the lesser apes, the great apes do not brachiate. The orangutan walks on branches while holding on to branches above with its hands; gorillas and chimpanzees walk on the knuckles of their hands, and on the curled toes of their feet. The gorillas have better opposability and make greater manipulative use of their thumbs than do the other apes, but it is the chimpanzee that makes and uses tools.

Whereas the monkeys use color for purposes of display and threat, the apes use their size and, in the orangutan, the fatty deposits around their face. These "blinkers" contain reserves that are drawn upon in times of food shortage.

The vertebral column and the skull
All monkeys carry their slender bodies horizontally on four limbs and the chest is deeper front-to-back than it is across. But apes walk erect or semiupright. They have fewer vertebrae in their trunks which are wider than they are deep. They thus characterize the evolutionary trend toward the reduction of trunk length in the more recently evolved primates.

Instead of a flattening of the face the great apes have developed prognathic jaws. Such heavy jaws and high cranial capacity are associated with increased brain size and result in a heavy skull. The forward position and the weight of the skull make the head fall toward the chest. This feature and an enlarged crest which runs from front to back across the top of the head is characteristic of many great apes. The crest is twitched in displays of aggression, but its height also serves to attract females.

The long fringes of white fur are the dominant feature of the black-and-white colobus monkeys *(Colobus* spp.) of Africa, and have led them to be hunted for the fur trade. These forest-dwellers are leaf-eaters and have a complex digestive system to deal with foliage.

Brachiation, the method of movement under branches by swinging from hand to hand, has been perfected by the gibbons *(Hylobates* spp.). The males also use this swinging and dropping action as a territorial display to neighbors.

Reproduction

The female chimpanzee alone among the apes exhibits external genital swelling as an indication of estrus. Female ape reproductive behavior is characterized by light menstrual bleeding and the sexual cycles last about a month. Gestation lasts about 30 weeks in gibbons (compared to 27 weeks in the much larger baboons) and 38 weeks in gorillas and orangutans.

Gorillas are highly social animals and live in groups consisting of about six families. Each group is ruled by one large silver-backed male. Their life is one of complex interactions between the members of the group, the rules of which are taught at an early age and upheld by the dominant males. The two subspecies of gorillas *(Gorilla* spp.) live on the floor of the lowland and montane forests of central Africa. They rarely climb and then only when young. Gorillas sleep on the ground on nests constructed from flattened foliage.

Homo sapiens

Evidence from fossils has convinced most scientists that human beings developed over millions of years from ancestors that were not completely human. The fossil record does not, however, provide enough information to trace human development in detail. As a result, not all experts agree on how humans developed. Man is, however, classified by scientists as a primate, and a survey of the living primates reveals that humans are most closely related to the apes and resemble them in a number of ways. Those trends that are preeminent in the evolution of the primates, such as the development of grasping hands, good stereoscopic eyesight, elaboration of the cerebral cortex and other parts of the brain, and longevity, are far advanced in the modern apes but have progressed furthest in human beings. But humans are vastly different from the apes in that their hands no longer perform a locomotory function, the brain is more than twice the size of that of any other living primate, and humans are bipedal — that is, they walk on two feet.

Hominid ancestors

Hominid fossils so far recovered from paleontological sites round the world have been classified into two genera, *Australopithecus* and *Homo.* The australopithecines of southern and eastern Africa lived from about four to one million years ago. Gracile australopithecus *(A. africanus)* was about the same size as the modern male chimpanzee and had a similar cranial capacity of 24 cubic

In an African tropical forest the layers occupied by different primate species are discrete. Boundaries are observed carefully to avoid confrontations and to reduce competition for the same food. The brown-and-red colobus *(C. badius)* feeds in the upper layer of the forest, which reaches about 150 feet (45 meters). The black-and-white colobus *(C. polykomos),* however, feeds in the middle story, at heights of about 100 feet (30 meters). Pottos and bushbabies eat among the upper branches of the under story, whereas the patas monkeys feed in the lower trees and on the ground. Gorillas inhabit the floor of the denser parts of the forest and eat ground vegetation as do the baboons, which live in the open savanna.

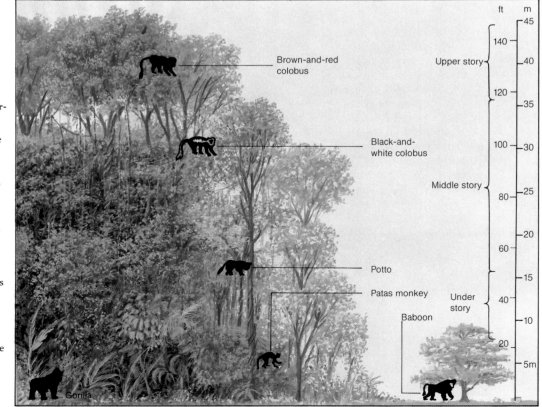

inches (400 cubic centimeters). In australopithecines the hole in the skull through which the spinal cord runs up to the brain (the foramen magnum) is relatively far forward and vertical. This suggests that the spinal cord must have entered the brain perpendicularly and the skull on the vertebral column would have faced forward rather than slightly downward, in contrast to the great apes. The australopithecine pelvis is bowl-shaped with short, stout hip bones which extend backward; their high surface area permitted the attachment of powerful gluteal (buttock) muscles and their long iliac crests allowed the anchorage of strong abdominal and back muscles. In these and other respects, many of their skeletal elements approximate those of modern man; the australopithecines were undoubtedly bipedal and walked upright.

Other hominids — *Homo habilis,* for example — shared parts of Africa with the australopithecines. But whereas the australopithecines may have used very simple tools — usually stone fragments — which they found but did not make, *Homo habilis* the tool maker definitely made them. With his somewhat superior cranial capacity of about 30—50 cubic inches (500—800 cubic centimeters) and less massive jaws, *Homo habilis* resembled modern man more than did the australopithecines. *Australopithecus* became extinct about one million years ago; *Homo habilis* may have overlapped with *Homo erectus,* modern man's immediate predecessor at about 1.5 million years ago. These early bipedal hominids were neither very large nor particularly strong, nor could they run fast. It is very likely then that cooperation based on communication and sign language, in food-gathering and group defense, would have been vital for their survival.

Homo erectus had, except for *H. sapiens,* the largest brain of all the primates, accommodated in a cranium with a capacity of 42—75 cubic inches (700—1,250 cubic centimeters). Despite the evolutionary tendency within the primates for the skull to become dome-shaped with increasing cephalization, *H. erectus* still had a sloping forehead with heavy, bony brow ridges and jutting jaws without a chin. But the brain's external casing has little to do with its efficiency.

H. erectus ushered in Acheulian stone technology which featured a wide range of carefully manufactured tools for the killing or butchering of game, and preparation of food. He was perhaps a good hunter: bone deformations caused by his excessive vitamin A intake as a result of eating too much raw meat have been diagnosed in some fossil remains. Fossil remains of this hominid, in East Africa, Europe, China and Java, point to his northward and eastward expansion out of his native Africa.

Modern *Homo sapiens*

According to scientific theory, modern man, *Homo sapiens,* began to emerge about 300,000

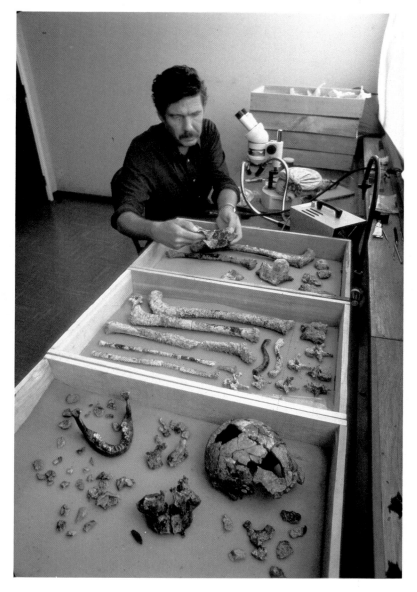

years ago. His earlier fossil remains have been found across Europe into Asia. He gradually spread from western and eastern Eurasia to America and Australia, helped by land bridges created by lower water levels during cold world climates. Human beings, at this stage, basically resembled *Homo erectus* but had a larger brain and smaller jaws and teeth. As time passed, *Homo sapiens* began to look like today's human beings.

Although the *Homo sapiens* face displays a highly differentiated musculature, man, nevertheless, relies primarily on symbolic language to express and communicate most aspects of his culture. Among primates, this characteristic is unique. Certain animals, including apes and monkeys, communicate by making a wide variety of sounds. These sounds express emotion and may communicate simple messages, but they apparently do not symbolize any object or idea. Language therefore distinguishes human culture from all forms of animal culture.

Bones of *Homo erectus* (above) were found in 1974 on the western shore of Lake Rudolf (Lake Turkana) in Kenya, Africa. Discovered between two layers of volcanic ash, the remains were dated by scientists as being 1.6 million years old, making them among the oldest *Homo erectus* finds. Because the skeletal remains are so nearly complete, have not been chewed by scavenging animals, and display no evidence of disease, scientists feel that future study of this find will provide valuable insight into the anatomy, growth, and development of these early hominids.

Rodents and lagomorphs

The order Rodentia is the largest mammalian order and is divided into three suborders: Sciuromorpha, which contains the squirrel-like rodents; Myomorpha, or mouselike rodents; and Hystricomorpha, or porcupinelike rodents. The word rodent means "gnawing animal"; these mammals were perhaps so named because their large incisors and their manner of eating are their most obvious characteristics. Lagomorphs were once thought to constitute a suborder of Rodentia because of their long, large incisors but they are now treated as a separate order (Lagomorpha). The order contains two families: the Leporidae, the hares and rabbits, of which there are some 50 species; and the Ochotonidae — pikas, or mousehares — with 14 species. Rodents and lagomorphs are similar in appearance and habits, but differ in certain aspects of anatomy. They are all relatively small animals and are found in most parts of the world.

General features of rodents

The two long pairs of chisel-shaped incisors in each jaw, characteristic of all rodents, project from the mouth and are used to grind down hard foods such as nuts and wood. These teeth, which are segments of a circle, grow continuously and must be constantly worn down at the tips. Rodents have no canines.

Because these animals feed on tough materials that are difficult to break down, they have a highly developed cecum (a branch of the gut at the junction of the small and large intestines), which aids digestion. A bacteria culture is maintained in the stomach which breaks down cellulose, which can then be absorbed by the stomach.

The limbs of rodents are constructed for plantigrade locomotion. There are usually five digits on the forelimbs and three to five on the hindlimbs. The digits on the forelimbs are extremely manipulative and are used for holding food. Hindlimbs may be adapted for running, jumping, climbing, or swimming (in which case the three to five toes are webbed).

The females are capable of bearing numerous young and usually do so at one time; some species can reproduce almost uninterrupted throughout the year. The uterus is usually divided, although in some species it is double. Most newborn rodents emerge both blind and naked and are entirely dependent on the mother for several days. But the guinea pigs (Caviidae), spiny mice (*Acomys* spp.) and coypus (*Myocastor* spp.) have precocious young which are active at birth and can soon take care of themselves.

Squirrellike rodents

Squirrels, marmots, gophers and beavers make up the suborder Sciuromorpha, in which there are seven families. These animals are found everywhere except in Australasia, Madagascar and parts of South America. They have a variety of habitats: some, such as the squirrels, are arboreal; others, such as the mountain beaver (*Aplodontia rufa*) and gophers, burrow underground; still others live a semiaquatic existence, such as beavers (Castoridae).

The squirrels (Sciuridae) are particularly noticeable for their bushy tails, and large eyes and ears. They are agile and graceful, and move equally well on the ground and in trees. The ground squirrels, such as prairie dogs (*Cynomys ludovicianus*), generally make complicated underground tunnels and chambers, some of which extend for hundreds of feet. The nest or burrows are sometimes used for storing food which is eaten when food is scarce. The marmots and woodchucks (*Marmota* spp.) are also ground-dwelling species of squirrels. Like the prairie dogs they live

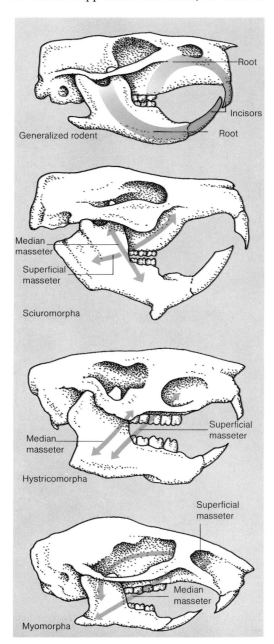

Generalized rodent
Root
Incisors
Root

Median masseter
Superficial masseter
Sciuromorpha

Median masseter
Superficial masseter
Hystricomorpha

Superficial masseter
Median masseter
Myomorpha

Rodent incisors grow continuously, from roots which extend deep into the jawbone. Rodents gnaw at any hard surface to keep the teeth in order. If their erosion is prevented, the tips of the teeth may grow past each other and those of the lower jaw may perforate the palate or grow out of the mouth into the eyes. The front of the incisors only has a hard coating of enamel which acts like a blade as the dentine behind it is worn down. The jaws that work these powerful incisors are operated, among others, by the masseter muscles. These muscles allow the lower jaw to pull back farther than in most mammals, enabling the cheek teeth to grind together. But they also bring the incisors together for gnawing.

in large social groups. They hibernate during the winter in their underground chambers and live off reserves of fat built up during summer feeding.

The beavers *(Castor* spp.) are the only members of the family Castoridae and are distributed throughout the Northern Hemisphere. Primarily water-dwelling rodents, they have dense underfur which is covered in coarse guard hairs. They are excellent swimmers and use their broad flat tails as a rudder while they paddle with their webbed feet.

The springhaas, or Cape jumping hare *(Pedetes capensis)*, is the only species of the family Pedetidae. In most respects it looks like a small version of the kangaroo, except that it has a bushy tail. Its long hind legs are used for jumping, but it can also move quadrupedally like a rabbit. Like the jumping hare, the kangaroo rats in the family Heteromyidae are nocturnal and burrowing, and their hindlimbs are also modified for jumping.

The brightly-colored scaly-tailed squirrels (Anomaluridae) are the only airborne rodents. They glide from tree to tree by means of a patagium, a membrane which joins their limbs.

Mouselike rodents

The suborder Myomorpha contains more than 1,000 species, arranged in nine families. Their mouselike appearance and lack of premolars distinguish them from the sciuromorphs. About 570 species belong to the family Cricetidae and include hamsters, lemmings, voles and New World rats and mice. They are mainly terrestrial, although some burrow and there are some semi-aquatic species. Many have a short tail and legs, and a thickset body.

The puna mouse *(Punomys lemminus)* is the only mammal to live at altitudes of 16,500 feet (5,000 meters). It is found in the altiplana region of the Andes and is about 6 inches (15 centimeters) in length. The more familiar common hamster *(Cricetus cricetus)*, which occurs across Europe into the Russian steppes, is a nocturnal species and may hibernate for a short period during winter. It stores food throughout the summer, some of which it carries to the winter burrow in its cheek pouches. This burrow is divided into separate compartments, each for a different food such as corn, or potatoes. They may store up to 200 pounds (90 kilograms) of food to see them through the winter. In summer another burrow is made which is used for nesting. The golden

hamster *(Mesocricetus auratus)* is found today only as a pet or laboratory animal.

Another member of this family, the true lemming *(Lemmus lemmus)*, lives in the Northern Hemisphere, particularly in Arctic regions. It is stocky, with very short ears and tail, and burrows in the soil during summer, and under the snow in winter. When food supplies are plentiful lemmings breed rapidly. But every four to five years their population reaches a peak when there are too many individuals for the available food. The lemmings then start to migrate in search of new food sources. In Norway, the migrating lemmings keep

The red tree squirrels *(Sciurus* spp.) of Europe, Asia and the Americas inhabit large conifer forests as well as mixed woodland and parks. The digits on their forelimbs are dextrous and used for holding food at which they gnaw. They learn to crack open nuts only through practice, and trial and error. Eventually, by chiseling deeply with their incisors into the natural grooves of the shell, they crack it easily.

The beavers *(Castor* spp., left) are known for their building techniques. They not only build complex lodges (below) but also engineer dams and cut streams through woodland. The dams and lodges consist of a foundation of mud and stones on which brush and poles are stacked and plastered with soggy vegetation and mud. The lodges are about 6 feet (2 meters) high with a diameter at the base of about 40 feet (12 meters).

Lodge   Ventilation shaft   Underwater entrance   Food store   Dam

Sleeping platform

Prairie dogs *(Cynomys ludovicianus)* inhabit the North American plains in underground colonies called towns. These towns can contain about 1,000 animals and are divided into communities of about 30 individuals. Sentries are posted at the entrance to the burrows where they keep watch with a characteristic stance, standing upright on their hindlimbs. If an intruder is spotted they bark a warning — the source of their name. They are known to clear the surrounding land of plants they do not like to allow those they prefer to grow.

moving until they reach the sea. This water barrier does not stop them and they plunge into the sea where most of them drown. In the following three to four years the numbers return to a similar level and the cycle continues. Lemmings compete for the same food as reindeer and caribou and, at the peak of their population numbers, they deprive the large ungulates of food and cause many of them to starve.

Some voles are also prone to similar cyclical fluctuations in numbers. The peak of population is known as a "vole year," during which there is a marked increase in the voles' predators, such as birds of prey.

Apart from the mole rats (Spalacidae), bamboo rats (Rhizomyidae) and jumping mice (Zapodidae), there are also the Old World rats and mice of the family Muridae. Many of them, especially the rats and the house mouse *(Mus musculus)*, have been introduced accidentally to all parts of the world, although they originated in the Old World. Most have long snouts and a long, naked, scaly tail. The black rat *(Rattus rattus)* probably originated in southeastern Asia, but like the house mouse, it is now found throughout the

world. Like other rats and mice, the black rat lives in close association with humans, inhabiting buildings and living off human food and refuse. These close associations often mean that the rats carry disease to humans.

The suborder also contains the dormice (family Gliridae) and spiny dormice (Platacanthomyidae). They are mainly arboreal and squirrellike with a bushy tail. Like other leaf-eating rodents they start to consume large amounts in autumn to accumulate as much fat as they can. This energy reserve enables them to hibernate throughout the winter, during which they may lose half their body weight.

Porcupinelike rodents

Sixteen families comprise the suborder Hystricomorpha, and include porcupines, guinea pigs and coypus. The porcupines make up two families: the Hystricidae, or Old World porcupines (found in Africa, Italy and southern Asia); and the Erethizontidae, the New World porcupines. In addition to hair on the body and tail, they have long, sharp, black-and-white quills on their back and flanks. The spines are loosely attached to the

Unlike most rodents of the temperate forests, dormice (family Gliridae) hibernate. This period of sluggish heartbeat and circulation can last from October to April and begins when the temperature drops consistently below 61°F (16°C). The dormouse's body temperature drops from its normal 98.6°F (37°C) to 39°F (4°C). It wakes occasionally when the outside temperature rises, to nibble at the food it has stored, and hibernates again when the temperature drops.

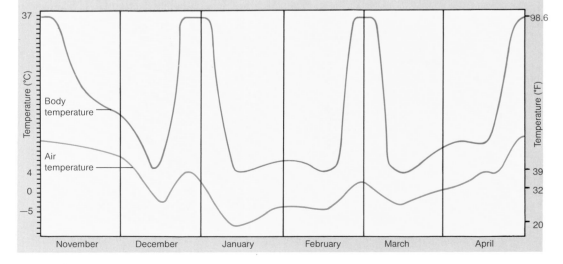

skin and when the rodent attacks, it lunges sideways, then immediately withdraws, so that some of the quills or spines may become lodged in the victim.

The New World porcupines have shorter spines and some of them are barbed. They are arboreal and their feet are modified for climbing. The sole is widened and the first toe on the hind foot is replaced by a flexible pad. This group is represented in South America, among others, by the family Caviidae — the cavies and guinea pigs — the capybaras (Hydrochoeridae), and the family Chinchillidae, the chinchillas and vizcachas.

Lagomorphs

The members of the order Lagomorpha differ from rodents in that they have an additional pair of sharp incisor teeth in the upper jaw, which are situated inside the normal pair. In addition these animals have a curious method of making the best use of plant food. To extract the maximum nutrient from their food, they first digest it in the stomach with the aid of bacteria and excrete it in the form of pellets. These pellets are eaten and pass through the digestive system once more — a process known as refection. After the second excretion they are left.

The family Ochotonidae consists of the pikas, or mousehares, found in North America and Asia. Of the Asian species, one lives on Mount Everest at an altitude of 5,780 feet (1,750 meters). Others live in deserts, on grasslands, rocky regions and in forests. Most of them inhabit areas with hard winters but they do not hibernate. They do, however, store food for winter and during the summer they dry vegetation in the sun, piling it up in "haystacks."

The remaining lagomorphs, the rabbits and hares, make up the family Leporidae. They are native to most countries except Australia and New Zealand although some have been introduced there by man. These animals are distinguished by their very long ears, short upturned tail and their hind legs which are usually longer than their forelimbs, allowing them to leap. Most are active at night or at twilight.

Hares (*Lepus* spp.) are solitary except in the breeding season. They do not burrow as rabbits do, but make a depression in the grass called a form, in which they rest during the day. They are most active after dusk, but are alert to potential danger. To escape predators they either run swiftly or lie prone and very still on the ground.

The European wild rabbit (*Orycolagus cuniculus*) has also been introduced to most countries. It is extremely prolific — females become fertile at about six months of age and usually have five to seven litters a year, each with three to seven young. They are capable of becoming pregnant within 12 to 15 hours after giving birth. They are also voracious feeders, and these two facts make them serious agricultural pests when they are not controlled.

Rabbits live together in large groups of about 150 individuals. They inhabit lowlands and hills, generally where the soil is sandy.

The Old World porcupines such as the Indian porcupine *(Hystrica indica)* have longer spines than their New World counterparts. This species has a short tail which, as well as being covered with sharp quills, has a cluster of hollow quills. These quills are shaken in warning and produce a sound similar to that of a rattlesnake. Porcupines attack in self-defense and usually run backward toward their attacker, releasing the quills in its face.

The hares *(Lepus* spp.) are notable mostly for their very long ears. Like rabbits, they eat bark, root crops and grasses but, in contrast to them, newborn hares are covered with hair, have developed eyesight and teeth. They are also solitary animals whereas rabbits are social.

Cetaceans

The cetaceans — whales, porpoises and dolphins — have the most highly evolved brain of all aquatic animals and include some of the most intelligent animals on earth; indeed, they are sometimes considered to be as intelligent as humans. Although they are descended from land-dwelling mammals, they have nevertheless become so well adapted for an aquatic life that they cannot now leave the water.

Except for a few species of dolphins that live in rivers and lakes (family Platanistidae), all cetaceans are marine mammals. The order Cetacea divides into two main types of whales, according to their method of feeding. The first type includes the toothed whales (suborder Odontoceti), which has seven families. The second type includes the baleen whales (suborder Mysticeti), which has three families.

The common naming of these animals is confusing; most large species are called whales and most of the smaller ones, dolphins. The killer whale *(Orcinus orca)*, however, is a member of the dolphin family (Delphinidae), and there are several species of whales considerably smaller than it.

A further inconsistency is that in the United States most dolphins are called porpoises, whereas in Europe this name is reserved for six species of small blunt-faced dolphins (family Phocaenidae).

The bottle-nosed dolphin *(Tursiops truncatus)* occurs in all the world's oceans, but is most common along the Atlantic coast of the United States. It eats fish and, like other true dolphins, can often be seen swimming alongside or in front of ships.

Anatomy and physiology

Whales and dolphins are the mammals best adapted for a life in water. The outline of their body has become very streamlined and the fore-limbs have become broad flippers used for steering and balance. These flippers still have all the bones of the vertebrate forelimb, although they have been modified. The hindlimbs have disappeared but a reduced pelvis is retained and supports (in the male) part of the reproductive apparatus. The dorsal fin and the tail are not supported by any skeletal elements, but are just folds of skin. The smooth outline is partly due to the disappearance of hair or fur. Propulsion is achieved by up-and-down movements of the broad horizontal tail flukes. The body is protected by layers of fat, or blubber, which act as an efficient insulator to keep the body warm, and as an energy store. The presence of large amounts of fat in the surface tissues allows subtle variations in the animal's shape which reduce drag and make deep diving easier.

Cetaceans usually show only a small part of their back when they come up to breathe, although even the largest ones can leap clear of the water. The nostrils are on the top of the head, forming a single blowhole in toothed whales, and a double blowhole in baleen whales. As a cetacean surfaces, it blows, sending up a distinctive spout of watery spray. This jet is a cloud of droplets caused by water vapor in the breath condensing as the pressure suddenly drops when the animal shoots up to the surface.

Cetaceans carry a relatively small proportion of their oxygen intake in their lungs, which are surprisingly small. There is evidence to suggest that the lungs are emptied before a deep dive and that oxygen is stored elsewhere, most of it combined with hemoglobin and myoglobin in the blood and tissues. Because there is not enough oxygen to supply all body functions during a long dive, cetaceans compensate by directing oxygen-rich blood towards the brain and nerves, where it is most needed.

These animals have also evolved some muscles that can work for short periods of time anaerobically (without oxygen), although they must be replenished with oxygen soon afterwards. The problem of nitrogen coming out of solution as bubbles in the blood when these animals return to the surface is reduced because of the small amount of oxygen they take down with them, and also because the rate of blood flow to the tissues is diminished.

A baleen whale filters krill from seawater using baleen plates that hang down from the palate, shown in the section of the jaws (right, A). An example is the blue whale *(Sibbaldus musculus,* far right, B); It is the largest mammal that has ever lived, although now in danger of extinction.

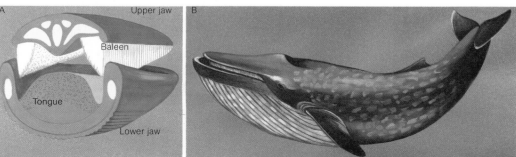

Cetaceans have very little sense of smell, and a variable ability to taste. Many species have good eyesight, but vision is not of much use because little light penetrates the depths of the sea. Sound, however, travels well in water and hearing is their main sense, as well as that of touch.

Whales and dolphins are able to produce a range of different noises, with which they communicate and find their way. Schools of dolphins chatter continually to each other, and the "songs" of the humpback whale *(Megaptera novaeangliae)* are known to carry hundreds of miles. Such calls inform the cetaceans of the identity and mood of their fellows. Distress calls, for instance, seem to summon other cetaceans to an injured individual.

Some species of toothed whales, especially dolphins, are known to use echo-location. Like bats, they emit streams of high-frequency clicks and ultrasonic squeaks and use the returning echoes to locate prey and to orient themselves. There is evidence that dolphins and killer whales have a language, and an ability to reason, to learn and remember information.

Feeding and dentition

The whales of the suborder Mysticeti have no teeth. Instead, they have a series of baleen plates (comblike keratin) hanging down from the palate. Their food consists of swarms of tiny crustaceans known as krill, and also sometimes of small fish. Water containing the food is drawn into the whale's mouth and then squirted out through the baleen plates, which act as a filter, keeping behind the food.

Toothed whales include all non-baleen whales and have numerous conelike teeth. But some toothed whales, such as the sperm whales (family Physeteridae), have teeth in their lower jaw only, and the beaked and bottle-nosed whales (family Ziphiidae) usually have only one or two pairs of teeth showing, but even these may not erupt. The

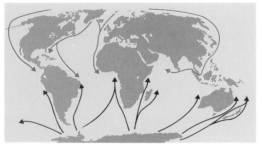

male narwhal *(Monodon monoceros)* has a single tooth which has become a long, twisted tusk on one side of its head.

The teeth of most toothed whales are adapted for holding prey but not for chewing; food is consequently swallowed whole. Most feed on fishes, squid and cuttlefish, but the killer whale will attack and eat any available prey.

Reproduction

Cetaceans breed seasonally, mating during the months between March and September. They produce single offspring and birth takes place about 11 months after conception in most species, although sperm whales have a very long gestation period of some 16 months.

Calves are born under water and may be as much as a third of the mother's size at birth. A calf emerges tail first and is pushed to the surface by its mother to take its first breath and to suckle from the teats, which lie in the folds of blubber on each side of the reproductive opening. The milk contains 50 per cent fat and is very concentrated.

Baleen whales seem to breed only every two years; as a result of this, the numbers already severely depleted by the commercial whaling industry will take a long time to recover. In addition, large whales start to reproduce only when they are about 12 years old, which obviously affects the replacement of individuals in nature.

A young humpback whale *(Megaptera novaeangliae)* swims above its mother. It is fed on mother's milk for five to ten months after birth, but does not grow to full size until it is ten years old. These whales migrate annually from Arctic and Antarctic waters (see map, left) to spend their respective winters in warm tropical seas.

Carnivores

The carnivores (order Carnivora) are the chief flesh-eating mammals, although hunting as a way of life is shared by others such as seals and some whales. But this characteristic of the group can be misleading, because a few species (such as some bears) do not eat flesh. The group, which contains seven families and more than 100 genera, is distributed throughout the world, apart from some islands and the Antarctic. Most carnivores are terrestrial or arboreal, but there are also aquatic species, such as the sea otter. Social patterns vary widely, according to the species, although most are fiercely territorial. The two most prominent groups, the cats and their relations, and the dogs and related species, have developed different hunting techniques — dogs, which are usually social and hunt in packs, tend to run down their prey, whereas most cats are solitary ambushers.

Carnivore anatomy

The long, flexible body of the hunting animals and their way of life are adapted to their predatory existence. Terrestrial carnivores are sure-footed and agile, as well as swift runners. Most species naturally stand on the tips of their toes with four or five toes touching the ground, and many have claws. Some species, such as dogs, cats and hyenas, have a dew-claw on each forefoot, which represents a toe that no longer touches the ground. The claws help to grip the ground and prey as well as being useful for digging and scratching.

One of the distinguishing features of carnivores is their long, pointed, canine teeth which they use for stabbing and holding prey. The skull of a carnivore is strong, and powerful muscles work the jaws. In some species there is a ridge of bone on top of the cranium (a sagittal crest) which serves to increase the anchorage for the upper end of the jaw muscles. The lower jaw is articulated only for open and shut action, with little facility for sideways grinding movements of the teeth.

In proportion to their body size, carnivores have a large and well-developed brain which controls the intelligent behavior needed for hunting. Their senses are efficient, enabling the animals first to locate prey at a distance and then to guide the attack. Carnivores have a fine sense of smell, important both for trailing prey and for communication. Scent released from the anal and other glands conveys information about an animal's identity, sex and territorial ownership. Vision is also vital for directing attacks, and these animals have forward-facing eyes which help in judging distance accurately when they spring on prey. The eyes of many carnivores have an internal reflector, the tapetum, which increases its sensitivity at night; it is the tapetum that causes the eyes to shine when a light is directed at them.

Most carnivores breed only once a year — although a few species reproduce twice a year, such as some weasels, and some breed only every two or three years. Litters vary in size from single offspring in bears to 12 or more in skunks. The young are usually blind at birth and are necessarily dependent on parental care for some time. In captivity some carnivores have been known to live for more than 30 years.

The flesh-eating mammals (Carnivora) are thought to have evolved from insectivore stock some 60 million years ago. By the end of the Eocene epoch two lines with distinct dog-like and cat-like characteristics had emerged, although fossils showing features of both have been found. The development of pinnipeds (walrus and seals) and their relationship with carnivores is difficult to ascertain because of the lack of fossil evidence.

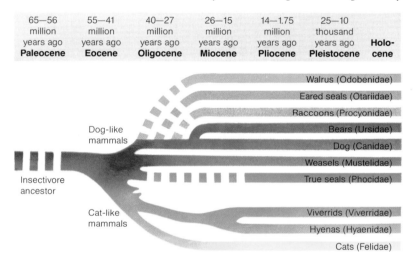

65—56 million years ago **Paleocene**	55—41 million years ago **Eocene**	40—27 million years ago **Oligocene**	26—15 million years ago **Miocene**	14—1.75 million years ago **Pliocene**	25—10 thousand years ago **Pleistocene**	**Holo-cene**

Walrus (Odobenidae)
Eared seals (Otariidae)
Raccoons (Procyonidae)
Bears (Ursidae)
Dog (Canidae)
Weasels (Mustelidae)
True seals (Phocidae)
Viverrids (Viverridae)
Hyenas (Hyaenidae)
Cats (Felidae)

Dog-like mammals

Cat-like mammals

Insectivore ancestor

Carnivores have four kinds of teeth: incisors, canines, premolars and molars. Incisors are for biting off flesh, canines for stabbing and grasping prey. The premolars and molars for slicing and chewing show considerable modification, according to diet. Bears chew their food thoroughly and have flat cheek teeth to crush vegetable matter. Cats and dogs have more pointed cheek teeth called carnassials, which slice and tear meat into chunks.

Typical bear

Molars

Premolars Canines

Incisors

Typical dog

Canines

Carnassials

Incisors

The dogs

The dogs (family Canidae) are generalized hunters. They are the most vocal of the carnivores, having a variety of barks, howls and whines. Those members of the family that hunt in packs can bring down large animals, but the solitary hunters usually live on small rodents, insects or birds.

The wolf *(Canis lupus)* looks like a heavily built German shepherd dog. Until persecution by humans drove it from much of its former range, it could be found over most of the Northern Hemisphere. The basic unit of wolf society is a female with her offspring. When the cubs mature, they stay with the mother and hunt with her. The male stays with his family and helps to feed the cubs. Solitary wolves eat anything they can obtain, from mice upward, but most wolves form packs of 2 to 10 animals, which are highly efficient hunting groups capable of bringing down prey as large as an elk.

The coyote *(Canis latrans)* resembles a small wolf. It is found throughout North and Central America and lives in pairs or family groups which feed on anything from deer to refuse.

There are about four species of jackals living throughout Africa and southern Asia, which usually live in pairs that cooperate in hunting, but they may gather in packs to prey on large animals or to scavenge.

The red fox *(Vulpes vulpes)* has as extensive a range as the wolf, and has been introduced into Australia as a quarry for hunters (an interference with the natural balance which Australians now regret). Unlike the wolf, the fox has survived continuous persecution and is even taking to living on the fringes of towns and cities, where it scavenges on domestic refuse. Some adult foxes are wanderers without a permanent home, but most live in pairs in territories, although they hunt separately.

The Arctic fox *(Alopex lagopus)* has two forms of seasonal color change. It has either white fur in winter, which turns brown in summer, or a blue-grey winter coat which becomes darker in summer. This variation seem to reflect the harsh-

ness of the climate and the environment.

The hot, dry parts of Africa are the home of the bat-eared fox *(Otocyon megalotis)* and the fennec *(Fennecus zerda)*. Both species have outsize ears which act as radiators to help to keep the animal cool, and as sound receivers to increase the sensitivity of hearing.

The Cape hunting dog *(Lycaon pictus)* of the African savannah is exceptionally social. It lives in packs of up to 20 that cooperate to hunt zebras and antelopes. The members of the pack share the food and, when a female has pups, the other dogs bring food back to the den.

The bears

The bears (family Ursidae) are the largest members of the carnivore order, and are found in the Northern Hemisphere and in parts of South

The foxlike fennec *(Fennecus zerda)* lives in the African desert, feeding on a varied diet of rodents, reptiles and insects; it also eats fruit, such as dates. It rarely drinks, but relies instead on the body fluids of its prey to provide it with moisture.

Cape hunting dogs *(Lycaon pictus)* live in the open on the grasslands of southern Africa, forming packs to hunt herbivores such as wildebeest and antelopes.

Bears are grouped with the carnivores and they do sometimes eat meat when it is available, like this Alaskan Kodiak bear, a type of brown bear *(Ursus arctos)*, which has caught a salmon. Most of their diet, however, consists of fruit and grasses, and so they can properly be regarded as omnivorous.

The hindlimbs of a dog (below) and a bear (below, right) reflect their ways of walking. Dogs are digitigrade: they walk on their fleshy toe pads. Bears are plantigrade: they walk on the sole of the foot with the heel in contact with the ground. A dog has blunt claws on its four toes, whereas a bear has sharp curved claws on each of its five toes.

America. They are heavy-bodied and most species live in forests; those that live on mountains have probably been forced there by humans. Bears are usually omnivorous and have blunt teeth. They eat a variety of plant foods, such as fruit and grass, but fish and other small animals are also eaten when they are available. Bears are solitary animals, although several may gather around a good source of food such as a rubbish tip or a shallow salmon-run. The cubs are extremely small at birth, weighing less than 1/350 of their mother's weight. The species that live in cold places spend most of the winter asleep in a den, but do not really hibernate.

The brown bear *(Ursus arctos)* weighs up to 1,700 pounds (780 kilograms). It lives in the northern forested areas of North America, Europe and Asia. The grizzly bear is a large subspecies of the brown bear, but is less common. Another North American bear is the black bear *(Euarctos americanus)*, which is smaller than the brown bear; it is an agile climber even when adult. Other bears include the spectacled bear (the only true bear in the Southern Hemisphere), moon bear or Asiatic black bear *(Selenarctos thibetanus)*, which is characterized by a white mark on its chest, and the sloth bear *(Melursus ursinus)* of India and Sri Lanka, which has a long snout with mobile lips which it uses to suck out termites

from their nests. The female carries her young on her back when they are first ready to leave the den. But the smallest and most arboreal bear is the sun bear *(Helarctus malyanus)* of Burma, Malaysia and Borneo; another characteristic is its short coat, which is less shaggy than that of other bears.

The polar bear *(Thalarctos maritimus)* is different from other bears, being forced by its habitat to be almost exclusively a flesh-eater. It lives in the Arctic and spends most of its time on pack ice, where it preys mainly on seals. It is a strong swimmer but never hunts in water; seals are either pulled out of the water (as are fish) or caught on the ice. Birds, hares and berries are also eaten at times. The thick white fur of the polar bear covers even the soles of its feet, giving added grip on the ice as well as insulation from the cold.

The raccoon family

The raccoon family (Procyonidae) includes the raccoons, cacomistles, kinkajous, olingos and pandas. Most are small animals which spend much of their time in trees. Apart from the two pandas, they live in temperate and tropical areas of the Americas and eastern Asia.

Most procyonids have long ringed tails; that of the arboreal kinkajou *(Potos flavus)* is prehensile. Their diet includes a variety of small animals, although the kinkajou and olingo *(Bassaricyon gabbii)* eat mainly fruit. Most procyonids are nocturnal, except for coatis *(Nasua* spp.*)*, which are usually diurnal. Most are also found in pairs or groups, except for the giant panda *(Ailuropoda melanoleuca)*, which is solitary. The coatis are the most social of this group, living in large bands.

The raccoon *(Procyon lotor)* is found throughout North America and in parts of Central America. Like the fox, it has adapted to living in built-up areas and scavenges in trash cans, but in wilder regions frequently feeds along the water's edge on shellfish, crayfish and frogs.

The question as to whether the giant panda is a near relative of the bears or of the raccoons remains debated. Answering it has not been made easier by the remoteness of the animal's habitat in the bamboo and rhododendron forests of western China. Its diet is almost exclusively bamboo shoots, and a feature characteristic of pandas is a sesamoid bone which rises from the wrist and acts as a sixth finger on the fore paw, used to hold the shoots. Pandas have broad teeth for chewing tough shoots into a pulp, and a muscular stomach for aiding digestion. The lesser or red panda *(Ailurus fulgens)* also lives in bamboo forests, from China to Nepal. It has a similar sixth finger for handling bamboo but frequently also eats acorns, roots, fruit and small birds.

The weasel family

The family Mustelidae, with 67 species, is the largest and most varied of the carnivores. Most mustelids resemble the European weasel *(Mustela nivalis)*, with long, slender bodies and short legs, and are unspecialized hunters of warm-blooded prey. From this basic pattern they have developed

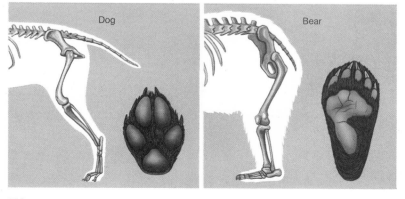

Dog

Bear

a variety of life styles: the martens *(Martes* spp.) are tree-climbers; the minks and otters are aquatic; and the badgers are stocky digging animals. The diet also varies among species. For example, the tayra *(Tayra barbara)* of Central and South America eats fruit, small animals and birds; the larger wolverine or glutton *(Gulo gulo)* is a powerful scavenger of the high northern latitudes which can also kill deer and cattle weakened by hard weather.

Weasels are lithe, very fast-moving animals. The European weasel and the American least weasel *(Mustela rixosa)* are closely related (the latter is the smallest carnivore). Weasels are small enough to follow mice and voles down their burrows, so do not compete directly with their larger relation, the stoat *(M. erminea)*. In the northern parts of their range (North America, Europe, Asia and Indonesia) stoats and weasels grow a white coat in winter, which makes them inconspicuous in snow.

Minks and polecats are larger members of the family, and the black-footed ferret *(M. nigripes)* is one of the rarest carnivores. It lives in the North American prairies and preys almost exclusively on prairie dogs (a type of squirrel). Its numbers have declined because prairie dogs have become scarce, having been much reduced in numbers by farmers.

Most carnivores have anal glands which secrete a fluid used for marking territory. But the best-known example is the skunk, also a mustelid, which uses the fluid for defense. The glands have also been modified for the same purpose in the Asian stink badger *(Mydaus javanensis)*, the African striped weasel *(Poecilogale albinucha)* and the African zorille *(Ictonyx striatus)*.

The badger *(Meles meles)* of Europe and Asia is one of the few social mustelids — and one of the largest — and lives in family groups called clans. The clan territory centers on their burrows, or setts, which may form an extensive underground network. Badgers are nocturnal and feed mainly on earthworms and insects. The American badger *(Taxidea taxus)*, however, is more solitary and its diet is mainly rodents. The honey badger or ratel *(Mellivora capensis)* is known for its association with the honeyguide, a bird that attracts the ratel to wild bees' nests. The ratel opens the nest to devour the contents, and the honeyguide feeds on the scraps.

When a striped skunk *(Mephitis mephitis)* is threatened, it raises its tail as a warning (A). It arches its back and may raise itself on its front paws (B). If the threat persists, the skunk turns and ejects two jets of foul-smelling fluid from its anal glands (C).

The raccoon *(Procyon lotor)* is an opportunist, feeding in the wild on frogs and fish or on the fringes of towns scavenging in trash cans and garbage.

The stoat *(Mustela erminea)*, or ermine as it is often called when in its white winter coat, can run extremely fast, chasing down its prey of small rodents or even rabbits and squirrels.

Eggs form part of the diet of the banded mongoose *(Mungos mungo)* of Africa, together with insects, snails and mice; it also eats small reptiles and fruit. Mongooses roam in bands, taking temporary shelter in the abandoned nests or burrows of other animals.

ecological niches in Old World tropical regions. Many have spotted coats and ringed tails. Some, such as the genets *(Genetta* spp.) are agile, cat-like tree climbers, whereas the otter civet *(Osbornictis piscivora)* and crab-eating mongoose *(Herpestes urva)* hunt in water.

Viverrids prey mostly on small animals, although some also eat fruit. Mongooses are known to attack snakes, using their speed and agility to confuse the snake and avoid its counterattacks. They have been introduced to several parts of the world to destroy poisonous snakes and rats, but cause destruction among native wildlife and also raid poultry pens (mainly for eggs). Most viverrids are solitary and active by night, because although they are well-armed, they are small enough to be attacked by birds, such as eagles, which hunt by day. Some of the diurnal species live in troops and cooperate to watch for danger and drive away large predators.

Some of the largest mustelids are the otters, of which 18 species live in fresh water, but some river otters and the sea otter *(Enhydra lutris)* frequent the sea and shoreline. Otters swim with powerful undulations of their bodies and broad tails, and steer with their webbed feet. They can dive for several minutes in search of fish and other aquatic animals, closing their ears and nostrils while submerged. The sea otter, which lives off the western coast of North America, feeds on crabs, sea urchins and shellfish. It smashes them open by banging them on a stone which it carries on its chest while floating on its back. It also carries and nurses its young in this position. Unlike most marine animals, sea otters do not have a heavy insulating layer of fat; instead they rely on the protection of a layer of air trapped by their long soft fur.

The mongoose family

The mongooses and civets (family Viverridae) resemble weasels and, indeed, occupy similar

The hyenas

Hyenas (family Hyaenidae) have had a bad reputation as cowardly scavengers, but are now known to be fierce predators as well as eaters of carrion. The three species that live in Africa and southwestern Asia are strong runners and hunt in packs, chasing herds of antelopes or zebras until a victim can be caught and pulled down. Their powerful jaws and large sharp teeth can crunch even the largest bones. Hyenas have a big head and well-developed forelegs. They have a characteristic trot, but are also able to run at high speeds. Spotted hyenas *(Crocuta crocuta)* are as noisy as many dog species, making howling and "laughing" sounds.

The African aardwolf *(Proteles cristatus)* is a rare relative of the hyena and, in contrast, has weak jaws, small teeth, and lives mostly on termites and other insects.

The cats

The family Felidae consists of six genera and 36 species. The main genus *Felis,* with 25 species, includes the puma *(F. concolor),* the largest of the

Tiger *Panthera tigris*

Lion *Panthera leo*

Leopard
Panthera pardus

Young cheetahs *(Acinonyx jubatus)* feed on an antelope that has been run down and killed by their mother. The fastest of all mammals, cheetahs can run at speeds of up to 70 miles (110 kilometers) per hour over short distances and often chase their prey for 450 yards (415 meters) or more before springing on it.

genus; the ocelot *(F. pardalis)*; the serval *(F. serval)* and many species of smaller wild cats, as well as the domestic cat *(F. catus)*. The other main genera are *Lynx*, which contains four species including the caracal *(L. caracal)* and the bobcat *(L. rufus)*, and *Panthera*, which also has four species — the jaguar *(P. onca)*, the leopard *(P. pardus)*, the tiger *(P. tigris)* and the lion *(P. leo)*.

Compared with other carnivores, the cats have a short muzzle and a broad, rounded head. With the exception of the lion, the fur is soft and often marked with spots or stripes. Cats are specialized hunters having lithe, compact bodies and large, sharp, scissor-like cheek teeth, called carnassials. The whiskers are developed and cats have acute sight, hearing and sense of smell. There is a wide variation in the size and appearance of the members of this family; the smallest wild cats are much the same size as the domestic cat, whereas the largest species, the tiger, has a body length of up to 9.2 feet (2.8 meters), or 10.5 feet (3.2 meters) including the tail.

Most cats are nocturnal and all, except the lion, hunt alone. Lions live in prides of several females and their cubs, with one or more males. Several members of the pride usually hunt together. They lie in ambush or sneak forward slowly toward their prey until they are close enough to leap up and catch it, bringing it down with their claws and a bite that breaks the victim's neck.

Unlike most carnivores, cats have retractile claws which are extended to help to grasp and slash prey, and when not in use are retracted into sheaths. The exception is the cheetah *(Acinonyx jubatus)*; its blunt, non-retractile claws give it a good grip on the ground as it sprints at up to 68 miles (110 kilometers) per hour. Its light build and flexibility help it to turn fast and its long tail provides steering and balance.

Among the big cats, one feature of some species is the roar, which is heard at night. These cats — the lion, tiger, leopard and jaguar — have an additional ligament in the throat which is attached to the hyoid bone.

The enormous variety in size and build of the cats becomes evident when species are compared. The head and body of a lion is about 8 feet (2.4 meters) long, and a tiger is even larger at up to 9 feet (2.8 meters) long. The leopard, cheetah and puma are similar in size — 5 feet (1.5 meters) — whereas the lynx and European wild cat are slightly smaller. The body length of the fishing cat is about 28 inches (70 centimeters). With the exception of the cheetah, these cats stalk their prey or spring from ambush.

Cheetah *Acinonyx jubatus*

Lynx *Lynx canadensis*

Fishing cat *Felis viverrina*

Puma *Felis concolor*

European wild cat *Felis sylvestris*

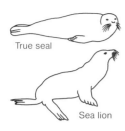

A true seal (above) cannot turn its hind flippers forward and so is much less mobile on land than the sea lion (below).

Pinnipeds

The seals (order Pinnipedia) are marine mammals which, together with the carnivores, insectivores, whales and humans, comprise the flesh-eating animals of this Earth. The order is divided into three families which, in turn, fall into two superfamilies — the true, or hair, seals (family Phocidae), the sea lions and fur seals, also known as the eared seals (family Otariidae) — and the walrus (family Odobenidae). Pinnipeds are found along most coasts but some of the biggest concentrations are in Arctic and Antarctic waters.

General features

Pinnipeds spend most of their lives in water, although they come on to land or ice to bear and rear their pups, and to molt. Like the cetaceans they are well adapted for an aquatic life; their bodies are streamlined and torpedo-shaped, and padded with fatty blubber which acts as an energy store and provides insulation, and the limbs have been modified into flippers for swimming. True seals are clumsy on land, whereas in water they are very graceful, swimming with side-to-side movements of the hind flippers, the foreflippers being used for maneuvering at low speed. Pinnipeds have either tiny external ears or none at all, and slit-like nostrils. The ears and nostrils are closed when the animals are submerged, but the eyes, which are well-cushioned, remain open and are efficient under water. Whereas walruses are almost bare-skinned, most pinnipeds have a covering of short hair.

All pinnipeds are well adapted for diving; when they plunge, the heart rate immediately drops from 50-100 beats per minute to 10 or less — the remaining bloodflow is directed mainly to the brain. Like whales, they can survive without breathing for much longer periods than land mammals; the record for diving is probably held by the Weddell seal (*Leptonychotes weddelli*), at a depth of some 2,000 feet (600 meters) and for up to 70 minutes.

An adult pinniped usually eats about 11 to 15 pounds (5 to 7 kilograms) of food per day. The diet consists mainly of fish but eared seals also eat crustaceans and marine mollusks. Others, such as the leopard seals (*Hydrurga leptonyx*), also eat seal pups, and a genus of sea lion (*Neophoca*) occasionally eats penguins.

About two weeks after a cow seal has given birth she mates again, but the implantation is delayed for a few months; this allows the female to recover from the strain of feeding her pup. It also means that pups are always born at about the same time of year, which is an important consideration for colonial or migratory species. Cow seals give birth to one pup a year, usually at a traditional breeding place where large colonies of seals gather. The bull seals at these breeding grounds are usually polygynous — one bull mates with several cows — and arrive there before the cows do to establish territories, keeping out other males by fighting.

After a gestation period of eight to ten months the pup is born. True seals suckle their pups for a short time — the harp seal (*Phoca groenlandica*) for example, suckles its young for only nine days after birth. The pups develop very rapidly and during this period the cow does not feed, drawing on her reserves of blubber to produce milk. The pups of some species are able to swim within hours of birth, whereas others take several weeks. Eared seals, however, rear their pups very slowly — Galapagos fur seals (*Arctocephalus galapagoensis*), for example, are not weaned until they are two or three years old.

Male pinnipeds are much larger than the females: a large bull southern elephant seal (*Mirounga leonina*) weighs 7,900 pounds (3,600 kilograms), four times more than the cow. The life span varies according to species, from about 25 years to 40 years or more, in captivity.

Elephant seals (*Mirounga angustrirostris*) breed in large colonies. The females (cows) feed the young (pups) on milk, living themselves on reserves of body fat, while the males (bulls) aggressively defend their harem of females.

The hindlimbs of true seals are contained within the body; the small forelimbs are placed well forward. They move with difficulty on land, hunching onto the front flippers (A) and wriggling forward (B); they haul themselves up (C) and then collapse into a new position (D).

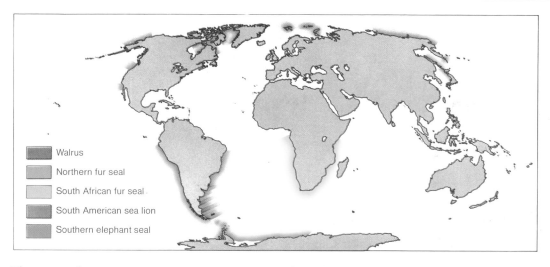

Walrus
Northern fur seal
South African fur seal
South American sea lion
Southern elephant seal

Some seals migrate. Walruses live in open Arctic waters, migrating south in winter before the ice closes in. Northern fur seals gather to breed at islands in the North Pacific, Okhotsk Sea and Sea of Japan, then disperse south. South African fur seals, South American sea lions and southern elephant seals are not really migratory, although the latter have been sighted as far north as Saint Helena.

The true seals

The 18 species of true seals live mostly in the Arctic and Antarctic, but some inhabit temperate waters. The main exceptions are the monk seals *(Monachus* spp.) which live in warm seas and some seals which inhabit freshwater lakes, such as the Baikal and Caspian seals *(Pusa subirica* and *P. caspica)* which live in these inland waters.

Pups of most species are born with a soft, dense, woollike white coat which is replaced by the stiffer adult coat before they start to swim. The white coat serves as camouflage on the ice and its thickness provides good insulation. Common seal pups *(Phoca vitulina),* however, are born with their adult coat.

The crabeater seal *(Lobodon carcinophagus)* is the most abundant of all true seal species, with a population of 15 million. Despite its name, it feeds on shrimplike crustaceans which it strains from the water through its teeth. Young crabeaters are hunted by leopard seals and many bear large scars from these attacks.

The eared seals

There are 16 species of eared seals, five of which are sea lions and 11 fur seals; they are found in polar, temperate and subtropical areas. They get their name because of their tiny cartilaginous ear-flaps, which true seals do not possess. They are long and slender, with short but developed tails, whereas true seals have only vestigial tails.

Fur seals have a thick undercoat of soft fur, protected by longer, coarser hair, whereas sea lions have one layer only of coarse hair. Sea lions are the biggest of the eared seals, have a heavier muzzle and head, and also a greater disparity in size between the sexes than other seals.

The eared seals and the walrus swim at high speeds by undulating their bodies, with their foreflippers tucked well back. The northern fur seal *(Callorhinus* sp.), for example, is able to swim at 10 miles (16 kilometers) per hour. Eared seals are more agile on land than are true seals because they can turn the hind flippers forward; they thus lift the body off the ground, and lollop along.

The walrus

The single species of walrus *(Odobenus rosmarus)* lives in the Arctic Ocean; it migrates south in winter, riding on the ice floes, and returns in the spring. It has a thick body and a tough, hairless skin which covers a layer of blubber 6 inches (15 centimeters) thick. Both sexes have a pair of tusks, which are long upper canines, to stir up food from the seabed, to fight, and to clamber up on to the ice. Walruses feed on bivalve shellfish. The bulls can measure 10 to 13 feet (3 to 4 meters) long and weigh as much as 3,000 pounds (1,401 kilograms). Despite their size they can swim at about 15 miles (24 kilometers) an hour.

Adult walruses *(Odobenus rosmarus)* have long powerful tusks — canine teeth — which the males brandish in threat displays when competing for females.

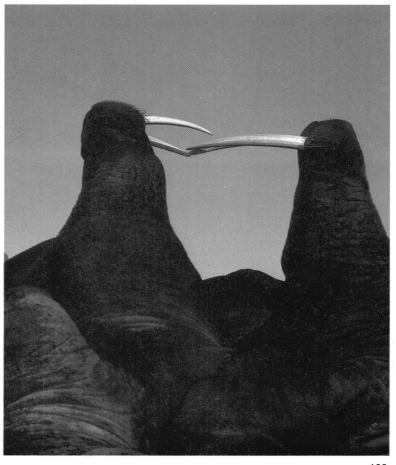

Aardvarks and subungulates

Indian elephant

Dugong

Rock hyrax

Aardvark

The subungulates, shown here to the same scale, have a wide variety of forms and sizes, but are classified together mainly because of their dentition. An aardvark is included for comparison.

Plant-eating mammals that possess hoofs are referred to as ungulates. But many non-hoofed animals have evolved from primitive ungulate ancestors and have similar features to the ungulates; consequently, they are called subungulates. These include the aardvark, elephants, manatees and hyraxes. Of these, the last three groups are herbivorous and have well-developed grinding molar teeth.

Aardvarks

As the only surviving species of primitive ungulates, zoologists place the aardvark *(Orycteropus afer)* in an order by itself — Order Tubulidentata. The name relates to the unusual molar teeth that the aardvark has, which contain many tubular pulp cavities. Aardvark means "earth pig" in Afrikaans, referring to its appearance and burrowing habits. These animals are stocky and powerful, weighing up to 140 pounds (64 kilograms). They have large ears which flatten to keep out the soil when they are burrowing, and which can move independently of one another. They have strong limbs and a thick, muscular tail. Aardvarks resemble the anteaters of South America in several ways, having an elongated snout, reduced teeth and a long, protrusible tongue. These features, however, reflect parallel adaptations to a similar diet, rather than a close phylogenetic relationship.

Aardvarks feed mainly on termites, using their strong claws to break open the sun-baked termite hills. They also collect the insects as they swarm across the ground. An aardvark's tongue can be protruded up to 18 inches (46 centimeters) and is covered with sticky saliva to which the insects adhere. Their thick skin appears to protect them from the painful bites of soldier termites.

Aardvarks are found in the savanna regions of Africa. They are nocturnal, traveling up to 19 miles (30 kilometers) a night in search of termites. During the day they sleep in their burrows, and so are rarely seen. They usually have single offspring, but occasionally produce two.

Elephants

The present order of elephants (Proboscidea) is a mere remnant of its former size. In the Ice Age numerous species, which included the mammoths, were spread throughout the world. Today, only one family remains (Elephantidae), containing two species — the African elephant *(Loxodonta africana)* and the Asian elephant *(Elephas maximus)*. They are easily distinguished from each other, because the African species has larger ears than its Asian counterpart and grows to a greater size. The African elephant is the largest living land animal, weighing up to 6 tons. To support their weight, the limbs of elephants have become pillarlike, with each bone resting directly on the one below. The feet contain elastic pads to cushion its weight as the elephant moves.

Elephants often consume more than 770 pounds (349 kilograms) of vegetable matter per day, and may spend from 18 to 20 hours a day feeding. Food and water are collected using the trunk, which is an extension of the nose and upper lip. At the trunk tip are finger-like projections (one in the Asia elephant and two in the African one) which allow the trunk to pick up objects as small as groundnuts. Most of their food is woody and fibrous, which is broken down by very large molar teeth with transverse ridges. Each jaw has six molar teeth per side, but normally only one tooth is present at a time. As the tooth wears away, another one erupts from the back of the jaw. Once the last tooth (the size of a small brick) is lost, elephants have difficulty feeding, and may starve. The much sought-after ivory tusks of elephants are not canines, as is often thought, but well developed upper incisors. Tusks of more than 10 feet (3 meters) long are known to have been taken from African bull elephants.

Hyraxes

Hyraxes, or "dassies" as they are often called, resemble large, grey-brown guinea pigs. But they are more closely related to elephants than to

Hyraxes are agile climbers, both on rocks and in trees. This rock hyrax keeps a lookout for possible predators, while the rest of the group are probably basking in the sun.

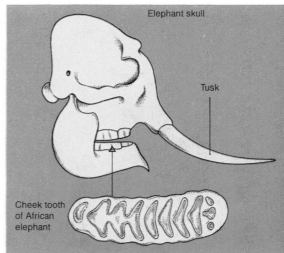

Elephant skull

Tusk

Cheek tooth of African elephant

rodents, particularly in the arrangement of their teeth. The members of this group (order Hyracoidea) have well developed grinding cheek teeth, and upper incisors which grow continually and curve, protruding from the mouth. These teeth are equivalent to tusks in elephants. The hyrax digestive system is unusual in that it has two appendixes.

Rock hyraxes *(Procavia* sp. and *Heterohyrax* sp.) are highly sociable animals and live in large colonies. They are diurnal and often bask in the sun, when older members of the colony act as lookouts, giving a shrill alarm call if a predatory animal, such as a leopard, bird of prey, or rock python, approaches. Tree hyraxes *(Dendrohyrax* spp.) are less sociable and are nocturnal, spending the day in tree holes or thick foliage.

All hyraxes are highly vocal and use a wide range of croaks, cries and alarm calls. All species also have a gland on the back, which is marked by a patch of different colored hair. When they are frightened, or during the mating season, this gland is exposed when the surrounding hairs are erected.

Manatees, dugongs and sea cows

Despite their large, blubber-filled bodies, this group (order Sirenia) are thought to have been the origin of the mermaid stories. These aquatic mammals are found in coastal waters and large river systems in tropical and subtropical areas. Weighing up to 1,500 pounds (680 kilograms) and ranging in size from about 6 to 14 feet (2 to 4 meters) long, they do not venture ashore, even when giving birth. Today there is one genus of manatees *(Trichechus)* which contains three species, and one genus of dugongs and sea cows (Dugong) of which only dugongs exist.

Sirenians are herbivorous, feeding on seaweeds or freshwater plants. In Guyana manatees have been used to control water weeds, which would otherwise choke up the rivers. The dugong has tusklike incisors and three cheek-teeth on each side whereas the manatee has no incisors, but up to ten cheek-teeth on each side which are replaced and moved forward continually.

African elephants roam in herds consisting mainly of females and juveniles. After a gestation period of nearly 23 months, pregnant females leave the herd to give birth. Like other mammals, the newborn young are fed on milk; the female has two nipples, located between her front legs.

The forelimbs of manatees form paddles, as in many other aquatic mammals. There are no hindlimbs, and the tail has a horizontal fluke like a whale's (but unlike seals and sea lions). Manatees can remain submerged for up to 15 minutes before having to surface to breathe.

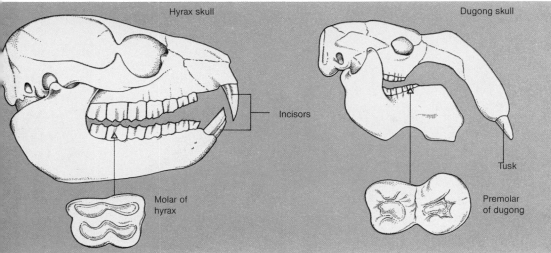

Hyrax skull

Dugong skull

Incisors

Molar of hyrax

Tusk

Premolar of dugong

Elephants, hyraxes and dugongs all eat plants and have batteries of cheek teeth for grinding plant material. A hyrax has rodent-like incisor teeth, but in elephants and dugongs the upper incisors are modified into tusks. A manatee has neither incisors nor tusks.

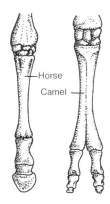

The horse represents the odd-toed ungulates (Perissodactyla), and the camel the even-toed ungulates (Artiodactyla).

The antlers of the red deer stag *(Cerbus elaphus)* are a typical feature of those herbivores whose dentition has been so modified for a herbivorous diet that they cannot use their teeth for protection or offense. Instead, these animals have developed horns or antlers with which they defend and assert themselves.

Ungulates

The ungulates are hoofed, herbivorous mammals, although some are omnivorous. The group is composed of several unrelated species of mammals that have evolved hoofs. They have been grouped together into two orders — the odd-toed hoofed mammals (Perissodactyla), which comprise about 15 species, and the even-toed ungulates (Artiodactyla), which consist of approximately 170 species.

The characteristics, common to all ungulates, are concerned either with the problems of extracting energy, during digestion, from the plant food they eat (a process called rumination), or with defense — they are the only horned animals, and those that do not have horns use their long canine teeth as weapons.

General features

The dentition of ungulates allows them to cope with a herbivorous diet, the constituents of which need to be thoroughly chewed. The sharp, pointed cheek-teeth found in primitive mammals are unsuitable for this diet and instead ungulates have developed large, flattened cheek-teeth with good grinding surfaces. In those animals whose diet consists of highly abrasive vegetation such as grass, molars with high crowns have evolved, which can be ground down a considerable distance before they are worn out. The cheek-teeth have complex ridges composed of dentine, which are adjacent to areas of enamel, and are filled with cement. As the vegetation is ground down, the cement, enamel and dentine are worn away to different degrees, producing a rough, self-sharpening surface.

Any animal unable to defend itself well against predators needs to be able to run fast, and most ungulates have evolved an efficient type of movement known as unguligrade locomotion. In fast-running animals, the upper bones in the limbs are short, but the lower bones are long, allowing lengthy strides. In addition, the bones in the feet are elongated, and by running on its toes the animal effectively adds a third functional segment to its limb. The toes themselves are lifted until they only touch the ground at the tips. Hoofs have developed, replacing the stability that the ungulates lost by not having the whole foot flat on the ground. These hoofs are composed of keratin, which makes them light, resilient and extremely tough.

Because the ungulates lift the backs of their feet to run on their toes the short side-toes cannot reach the ground; without a function these toes have become reduced in many ungulates or have disappeared altogether.

It is this feature that divides the ungulates into the two orders. In the odd-toed perissodactyls the functional axis of the leg passes through the third or middle digit, and species with only three

Dentition among the ungulates is varied. The grazing horse (A, *Equus caballus*) has high-crowned cheek teeth with self-sharpening grinding surfaces. The tapir *(Tapirus* spp.), which nibbles at low shrubs (B), has relatively low-crowned cheek teeth and its nose extends into a short mobile proboscis. Among the artiodactyls, the peccaries *(Tayassu* spp.) have tusklike upper canines (C) and a long snout. The camels *(Camelus* spp.) have spatulate incisors (D) which project forward.

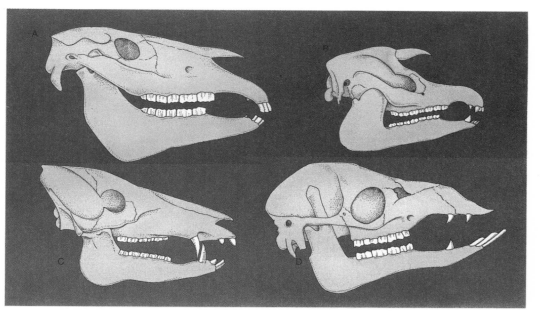

digits (or even just one) have evolved. Where the axis falls between the third and fourth digits, animals with either four or two toes have evolved — these are the even-toed artiodactyls.

Ungulates breed usually once a year, with single offspring, although some species breed once every two years. The females come into estrus several times a year and if not mated can continue on heat for several months. The gestation period is usually 11 to 12 months, and the life span from 25 to 40 years or more.

Rumination

Most mammals are not able to digest cellulose, but herbivores have evolved a method of extracting nutrition from their food by breaking down the cellulose. This is achieved by microorganisms which live symbiotically in some part of the digestive system and dissolve the cellulose.

In ruminating animals such as camels and chevrotains, the stomach has three compartments, but in true ruminants — giraffes, deer, antelopes, sheep, goats and cattle — it consists of four: the rumen, the reticulum, the omasum and the abomasum. Food, mixed with saliva, is fermented in the rumen and reticulum where bacteria and protozoa break it down. To ensure that the food is well broken-down, it is regurgitated and chewed once more (this is called chewing the cud). After it has been remasticated the food bypasses the rumen and enters the omasum. Here the mixture is fermented further; the liquid is pressed into the abomasum, where it is subjected to the action of digestive enzymes; the rest is absorbed and the waste is passed out.

Perissodactyls

The odd-toed ungulates have not been very successful. In the early days of mammalian history they were numerous, but today only three of the original twelve families remain: the horses (Equidae), the tapirs (Tapiridae) and the rhinoceroses (Rhinocerotidae). The tapirs and rhinoceroses are

greatly reduced in numbers and, apart from the zebra, the Equidae would probably also be greatly reduced in numbers had they not been domesticated. Their natural range today is Africa and the steppes of Asia.

The perissodactyls have three toes (if only on the hind feet, as in tapirs) or one single toe, as in the equids. They are browsers and grazers and the structure of their lips facilitates the collection of plant material. Their flat-topped cheek-teeth and high-crowned molars enable the breakdown of coarse vegetable food. They have simple stomachs, with a large cecum and no gall bladder. Horns composed entirely of dermal material may be present, and the skin is often very thick, with little hair.

The family Equidae contains only one living genus, *Equus,* which includes horses, the ass and zebras. All species are highly specialized for swift movement and for grazing. Only the third digit remains on the limb and the second and fourth

The warthogs *(Phacochoerus aethiopicus)* are typical of the hog-like artiodactyls in that their canines form tusks, which they use in defense. The nyala *(Tragelaphus angasi)* — like many other antelopes — has developed, instead, sharp horns with which it can defend itself.

The well-developed canines of the hippopotamuses are used as powerful weapons. These animals live in large herds, the males of which are territorial and use their lower canines in fights during the mating season.

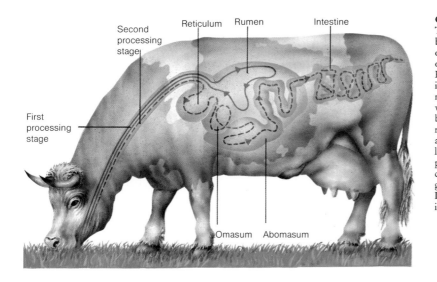

Second processing stage

Reticulum   Rumen   Intestine

First processing stage

Omasum   Abomasum

Cattle are true ruminants. They have a four-chambered stomach composed of a rumen, a reticulum, an omasum and an abomasum. Food is swallowed with salivary enzymes and is fermented in the rumen, where the enzymes and bacteria break it down. It is regurgitated and chewed again. The food is swallowed once more, this time passing into the other chambers, where it undergoes further enzyme action. It is absorbed and the waste is passed out.

toes have been reduced to splints. In their native states they live in migratory herds, as do many herbivores that live on plains.

The wild horse is now represented by only one species — the Mongolian wild horse or Przewalski's horse. They used to roam the steppes of central Asia in large herds but there have been no sightings of them for many years and they probably no longer exist outside captivity. The wild horse is shorter than the domestic one, with a stiff, erect mane, small ears, a low-slung tail and a shrill voice.

The ass, native to Africa and Asia, is found most frequently on plains sparsely covered with low shrubs. It was the first animal in the genus to be domesticated.

Zebras, now the most common member of the genus in the wild, are found only in eastern, central and southern Africa. They were once called horse tigers because of their stripes. The stripes of the plains zebra are wide apart, with shadow stripes in some species. The rare mountain zebra has wide stripes with a transverse gridiron pattern on the lower back. The largest zebra, Grevy's zebra, has narrow, close-set stripes.

The tapir, whose natural range is restricted to the Malay Peninsula, Java, Sumatra and Central and South America, has only one living genus *(Tapirus)* with four species. The tapir's forefeet have four toes, whereas the hind feet have three.

The four genera of rhinoceros are found in Africa, Asia, Sumatra, Java and Borneo. They inhabit savanna and moist wooded areas, often near water, feeding on shrubs, leaves and fruit,

and are mainly nocturnal. These massive animals — they reach 6 to 13 feet (2 to 4 meters) in length — have short, pillar-like limbs with three digits. They have thick, sparsely-haired skin with characteristic folds and one or two horns on the nasal plate composed of solid fibrous keratin. Their hearing and sense of smell are well developed, although their vision is dull.

Artiodactyls

In the early days of ungulate development artiodactyls were not as common as perissodactyls. But they have become increasingly prominent and today are one of the most successful groups of mammals, native throughout the world, except in Australia and New Zealand, where they have been introduced.

There are nine families of artiodactyls: the hog-like species contain the pigs and hogs (Suidae), the peccaries (Tayassuidae), and the hippopotamuses (Hippopotamidae). The camels and llamas are contained in the family Camelidae, and the chevrotains in the family Tragulidae. The rest of the artiodactyls are true ruminants and are grouped into the infraorder Pecora. This group contains the giraffes (Giraffidae), the deer and their allies (Cervidae), the cattle, antelopes, goats and sheep (Bovidae), and the pronghorn (Antilocapridae).

The hog-like artiodactyls are in many ways the most primitive of the group. They are still four-toed (although the side-toes are reduced), and their limbs are not greatly elongated, which means that they cannot move quickly. They are

The horse is the fastest, strongest runner of the ungulates, because in it the development of the unguligrade limb is most advanced. As a migratory animal it is adapted to sustaining speed for long periods.

The zebras *(Equus* spp.) are the most common equid in the wild. No two individual zebras are alike in the pattern of their stripes. It is thought that the stripes serve a purpose in social recognition. Out on the open plain they seek safety in numbers and move in vast herds of males and females. This characteristic contrasts with many of the other plains herd animals, such as antelope, which move in herds of separate sexes.

omnivorous and have large, canine tusks. Their two-chambered stomach is not as developed as that of the ruminants.

The most typical of the five genera of wild hogs is the wild boar *(Sus scrofa)*, from which the domestic hog was probably derived. The hog is typically a nocturnal forest-dweller, and travels in groups of up to 50 individuals. But some species, such as the warthog *(Phacochoerus aethiopicus)*, are diurnal and travel singly or in family parties. The males have upper and lower canine tusks which curve upward. They are mainly vegetarian but occasionally scavenge on carrion. The young number up to 12 per litter.

Peccaries are closely related to hogs, although they are smaller and lighter, and the males' upper canine tusks curve downward. Their stomach is slightly more complex than the hog's, perhaps foreshadowing that of the ruminants. There is one genus *(Tayassu)* which is restricted to the New World, occurring in deserts, woodland and forests, in bands often reaching 100 or more in number.

There are two genera and two species of hippopotamuses, both found in Africa. These large, heavy animals are adapted for both aquatic and terrestrial life. They have a broad muzzle which is suitable for taking in large masses of pulpy water plants. Their eyes and ears are situated high up on the skull, enabling them to function well while almost totally submerged. They are expert swimmers and can stay submerged for more than five minutes. These animals live in herds of 5 to 30, usually near the water.

The camelids can be split into two groups: camels *(Camelus* spp.),which are native to the Old World, and llamas *(Lama* spp.), found in South America. There are two species of camels, both desert-dwellers, although only one is still found in the wild. They are able to conserve water by reducing evaporation and concentrating their urine, and can lose up to 25 per cent of their body weight by desiccation. Possibly because of this characteristic they are unique among mammals in that their red blood cells are oval rather than spherical.

The hoofs have disappeared on all camelids and have been replaced by a nail and a large pad, which enables them to walk on soft or sandy ground. They have two digits on each limb, the third and

The tapir *(Tapirus* sp.) is a solitary animal found in damp forests or swamps, and is nocturnal or crepuscular. It feeds on leaves, shoots and fruit which it collects with its short proboscis.

In a transverse gallop, such as this (below), a horse's body is supported by at least one limb on the ground, except during one short phase in the cycle.

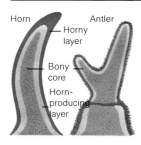

Horn	Antler
Horny layer	
Bony core	
Horn-producing layer	

Horns are a feature of most bovids. They are firmly attached to the skull and are not shed. In contrast the antlers of most cervids, which are bony growths from the skull covered by a layer of velvety skin, break off every year.

Camels *(Camelus* spp.) are well adapted for walking on sandy ground. Their digits (two on each limb) splay as they walk, revealing skin in between them which provides a larger surface area for them to walk on and helps them not to sink into the sand.

A stag's antlers are grown and shed every year, each time developing more points, until the animal reaches about its fifteenth year. They are shed in early spring and six weeks later new ones develop. By the end of May they are fully grown.

the fourth. The camelids are ruminants but have a less complex system than the Pecoras with no separation of the omasum and abomasum.

The South American camelids live in a variety of habitats, from cool plains to mountains of permanent snow. The llamas *(Lama peruana)* are smaller than camels, have no humps, and are covered with a thick coat of wool. Two wild species of South American camelid are the vicuna *(Vicugna* sp.) and the guanaco *(Lama guanacoe).*

The chevrotains, or mouse deer, are timid forest browsers, not much bigger than a rabbit, found in the Old World tropics. They have no horns and use their large, tusklike upper canines as weapons. They resemble deer in the white patterning of their red-brown coat. With a three-chambered stomach, the chevrotains — like the camels — lie between the non-ruminants and the true ruminants.

The rest of the artiodactyls, all of which are true ruminants, contain the most successful and numerous of the ungulates. As well as the posses-

sion of a rumen almost all of them have horns or antlers, and the side-toes of their limbs have disappeared, leaving two functional digits, although lateral ones may be represented by imperfect dew claws. The incisors have been lost from the upper jaw and the lower teeth bite against a strong horny palate.

The giraffes *(Giraffa* spp.) and okapis *(Okapia* spp.) are browsing animals now restricted to tropical Africa. Their long necks contain seven vertebrae only, which is the usual number in mammals, but each vertebra is greatly elongated. The neck of the okapis is shorter than that of the giraffes. Horns are found in both sexes and are ossified knobs which grow continually although slowly. Whereas okapis are solitary, giraffes usually move in family herds.

The deer, or cervids, which are mainly forest-dwellers, are essentially temperate zone animals although some are found in southern Asia. The males have antlers, which are bony growths from the skull. They are shed each year and form progressively more branches as the animals age. The older the cervid, the larger are the antlers.

Cervids are divided into four subfamilies: Moschinae, Muntiacinae, Cervinae and Odocoileini. Moschinae contains only one genus, the musk deer *(Moschus* spp.). Restricted to Asia, this genus is thought to be the most primitive of the cervids and, with the water deer *(Hydropotes* spp.), is the only type with no antlers. The male has an abdominal gland which exudes musk. Because of the demand for musk by the perfume trade resulting in it being extensively hunted, the musk deer has died out of most of its former range.

Muntiacinae, or muntjacs, contains two genera which are also restricted to Asia. They are small deer and although they have antlers they also have well-developed upper canines. They are solitary crepuscular animals.

The subfamily Cervinae includes the red deer and wapitis *(Cervus* spp.), axis deer *(Axis* spp.) and fallow deer *(Dama* spp.). They are found in Europe and North America. The Odocoileini is

Eighteen months old	Two and a half years old	Four and a half years old	Fifteen and a half years old
First set	Second set	Fourth set	Fifteenth set

Giraffes *(Giraffa* spp.) are the tallest living land mammals, the males reaching a height of 18 feet (5.5 meters) and the females 15 feet (4.5 meters). Because rising from a lying position is so difficult for these animals, they usually sleep while standing. They have massive hearts, needed to force the blood up their long necks to the brain. An advantage of their height is that they have little competition for the leaves on high branches on which they feed.

the most widely distributed subfamily and includes such species as white-tailed deer *(Odocoileus* spp.), moose or elk *(Alces alces)*, and caribou and reindeer *(Rangifer tarandus)*. The caribou and reindeer are the most northern species of cervid.

The bovids are the largest group of ungulates found throughout the world. They include most of the animals that man depends on for food, such as cattle, sheep and goats, as well as other animals such as antelopes and gazelles. There are nine subfamilies: the cow-like Bovinae, the duikers (Cephalophinae), the reedbuck and their relatives (Hippotraginae), antelopes and gazelles (Antilopinae), and the sheep and goats (Caprinae).

All bovids have bony horns which, unlike antlers, are never shed. The horns, which are firmly attached to the skin, are not branched, although they can be compressed or twisted. Bovids are usually found in herds and most inhabit grassland, scrubland or deserts, but goats and sheep are generally found in rocky mountainous or desert

areas. They feed by twisting vegetation around the tongue and cutting it with the lower incisors.

Domestic cattle, sheep and goats have been bred from Eurasian species. There are six species of wild sheep found in North Africa, Canada and mountainous areas of Asia and the Mediterranean. None has as woolly a coat as the domestic sheep. Together with rats and pigs, wild goats or goats that have escaped domestic captivity are among the most destructive of animals because they eat anything in sight. In destroying vegetation they have been instrumental in the extinction of other animal species, such as some birds.

The pronghorn *(Antilocapra americana)* is the only remaining species of its family, peculiar to North America. Herds of about a thousand animals used to be common but their numbers are greatly reduced and today they live in herds of about 50 to 100. Like cattle, their horns consist of a bony core which is never shed, although the horns' covering is. The horns are branched, as are those of the cervids.

Impalas *(Aepyceros melampus)* gather in large herds, often in groups of hundreds of individuals, which affords them some safety from predators; while a few sentries keep watch, the others can eat reasonably peaceably. Impalas can run very fast when threatened and are known to make successive leaps into the air to a height of about 26 feet (8 meters).

Animals in danger

A large flightless bird, the great auk *(Pinguinus impennis)*, lived in similar habitats in the Northern Hemisphere to those occupied by penguins in the South. It became extinct in about 1845; now only stuffed museum specimens remain.

The North American bison *(Bison bison)*, or buffalo as it is also known, was relentlessly hunted throughout the nineteenth century so that by the 1880s its numbers had fallen alarmingly. The animal has been brought back from the edge of extinction by careful management, but like its European cousin the wisent *(Bison bonasus)*, is today found only on reserves.

In a world that did not change physically and in which new species of animals and plants did not evolve, extinction would be unknown. Each available niche in the world's ecology would be filled by a steady population of various animals — the predators and the predated — in an unchanging, balanced life cycle. The real world does not correspond to these ideal conditions, however, in a number of significant respects. The population of a given species increases in a period when conditions are advantageous, and decreases when adverse conditions predominate (for example, when there is a particularly cold or long winter). The natural ecological balances that do exist are dynamic in character, and tend to change with time. In fact such changes are an inextricable part of the mechanism of evolution: if evolution takes place, extinction is inevitable.

Physical changes and extinction

The conditions for the existence of life on earth are dependent on a narrow band of acceptable criteria and only small variations in such factors as the local rainfall, temperature range or chemical pollution levels can lead to the extinction of a variety or a species. And this extinction in turn inevitably has some effect on all the local species as the eco-balance readjusts to compensate and regain stability.

A larger climatic or other change can have even more dramatic effects. For example, strong evidence now exists to show that the sudden and total disappearance about 65 million years ago of the dinosaurs (then the dominant life-form) and many other species resulted from catastrophic climatic changes caused by a collision between the earth and a large meteor or comet. Again, periodic cooling of the Northern or Southern Hemisphere in the form of ice ages (which might have been caused by variations in the sun's activity) had a dramatic effect on flora and fauna — leading to the reduction or extinction of some species and the increase or evolution of other better adapted ones.

Effects of one species on another

The gradual evolution of better adapted life-forms causes the reduction in numbers and possibly the eventual extinction of less well adapted species trying to occupy the same ecological niche. In addition, certain species develop an interdependence on other species of flora or fauna in areas of reproduction, habitat, or food requirements. Any external event that affects the numbers or habits of one of these species may therefore have disastrous effects on the survival of the other dependent species. Such interdependence reduces the chances of survival of the species concerned. And disease — particularly the evolution of new virus strains — threatens certain species with extinction while leaving others relatively unaffected.

In each ecological environment one dominant species is likely to evolve, and it has a major influence on the ecological balance, affecting the numbers and survival of other species. This dominant species is most likely to be the predator at the top of the food chain. It will be challenged and replaced in this position only by the evolution of a better adapted species, or by a physical change in the environment.

Man and his activities

The evolution of humans as an intelligent and highly adaptive species has led to an ecology in which they have a position of dominance over all other species on earth. This fact, linked with human ability to modify or even totally alter the environment, has placed many species in danger of extinction. Pressures of food production for a rapidly expanding human population have meant that there is probably no natural habitat that is not now threatened. Humans are the first species to have the ability to make all life forms extinct — including themselves.

The threat of mankind to other species is apparent in a number of ways. The development of domesticated strains of plants and animals to improve the efficiency of food production (for example, wheat and domestic cattle) demand the creation of specialized environments which remove large areas of habitat from the native ecology. Indeed, positive action is taken to exclude the naturally occurring plants and animals from these areas with such things as weedkillers, traps and fences. Increased requirements for living accommodation remove land from the natural ecology as new towns and cities are built. Relocation of

water courses and the physical manipulation of the landscape remove further natural habitats. Also overgrazing of domestic animals can reduce previously fertile land to new desert, again changing the local ecological balance.

The overexploitation of a species for food, clothing or ornament can reduce its numbers to below that necessary to maintain a viable population. The collecting of animals from the wild for study or to populate zoos can threaten scarce animals. Wars between various human groups can result in animals confined to a small area being threatened as they become incidental casualties, whether of bomb, bullet, fire or from the use of defoliants — or more indirectly by the lack of food which often accompanies war.

Human activities such as farming and industry create pollution by releasing poisonous chemicals into the atmosphere, soil and water. Species with little tolerance to such pollutants may be threatened. Sometimes catastrophic accidents that release large amounts of oil, toxic gases or radioactive waste can have immediate and disastrous effects on wildlife. The burning of wood and fossil fuels to produce energy may have long-term effects on the carbon dioxide content of the atmosphere — possibly affecting the climate in an adverse way. Other experiments in climate control might have unpredictable effects on the world's ecology.

The introduction of species from different parts of the world into an ecology — either accidentally or deliberately — can have a disastrous effect on local species and may result in the extinction of those that are unable to compete or those that become predated upon.

Wherever a species is present in only a limited geographical area, or occupies only a particular type of habitat, or has only one kind of food, it must be considered to be in potential danger. But ironically for some species whose numbers have fallen below a certain level, only the direct intervention of humans can now save them from total extinction.

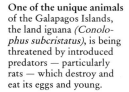

One of the unique animals of the Galapagos Islands, the land iguana *(Conolophus subcristatus)*, is being threatened by introduced predators — particularly rats — which destroy and eat its eggs and young.

Uncontrolled hunting and poaching still bring the threat of extinction even closer to some of the world's large animals. Despite international agreements, whales are still being overhunted. Fast catcherboats armed with explosive harpoons take the dead whales to large factory ships (above), where they are processed. In Africa, the killing can be even more wasteful. The rhinoceros is killed by poachers only for its valuable horn (left); the carcass is left for the vultures and hyenas.

Endangered lower animals

Increasingly people are becoming aware that various species of birds and mammals have become in danger of extinction, and commend or actually help in the steps being taken to conserve such animals. But many other creatures — the less glamorous animals such as mollusks and insects, fishes and reptiles — are also endangered. Colorful tropical butterflies are caught by the thousands so that their wings can be used in jewelry and ornaments. Some species of snails, crabs and crayfish are caught and eaten as delicacies in restaurants around the world without any thought about the preservation of breeding populations and their future survival. Fishes continue to suffer from the effects of overfishing and increasing pollution of the seas and oceans.

Nearly all such lower animals are part of the food supply of an interdependent chain of higher animals such as birds and mammals. Eventually some of these may also become threatened as their food supply diminishes. And ultimately the threat may extend to the food supplies of man himself. Yet it is nearly always the activities of man that have brought the animals to the point where their survival is in question.

Manmade hazards
The greatest of the threats to lower animals is pollution of all kinds — from the testing of nuclear weapons to the dumping of sewage and industrial waste. For example, the komodo dragon *(Varanus komodoensis)*, the largest of all the lizards, now survives on those tiny Indonesian islands that as yet offer nothing for man to exploit.

Land reclamation, industrialized agriculture and the wanton and greedy destruction of the world's large forests will continue to cause havoc among wildlife unless it is halted — and with some urgency. In some areas of the world there is little hope that present governments will realize that it is to the advantage of their own people and traditional ways of life that each and every species is conserved, including the lower life forms. An animal would not have evolved naturally if there was not some reason for its existence in the delicate interrelationships of nature.

In the sea
An instructive example of the threat to lower animals is provided by the myriad array of fishes and other creatures that depend for their existence on Australia's Great Barrier Reef. By the 1960s, the middle third of the coral reef had died off as a result of pollution and a plague of large, dark, spiny starfish known as the crown-of-thorns *(Acanthaster planci)*. These starfish, often as many as 15 of them on each square yard of reef, prey on the depleted coral polyps. The large increase in the number of crown-of-thorns is due in part to the removal from the reef of their chief natural enemies — and population stabilizers — the large marine snails. Called tritons (Cymatiidae) and helmet shells (Cassididae), their ornamental shells were much prized as souvenirs by tourists. The sale of the shells has now been halted to allow the coral to revive, so that a healthy balance of species is restored to the reef.

Concentrated fishing (and pollution) has resulted in shortages in the world's fish population. Since the 1940s so intensive has the fishing of some species become that even common fishes such as flounders, ocean perch, lake herring, tuna and others are presently in danger. Often the fault lies not merely with the tonnages of fish that are

The living corals whose skeletons accumulate to form the Australian Great Barrier Reef are the chief food of a spiny starfish (left) called the crown-of-thorns *(Acanthaster planci)*. The reef is a habitat for a whole range of interdependent creatures, which are therefore also threatened by the demise of the coral.

Butterflies have long been collected indiscriminately, mainly because they are so decorative. This New Guinea species, the Victoria birdwing *(Ornithoptera victoriae)*, has been hunted almost to extinction, as have other birdwing species and the iridescent blue South American butterflies such as those of the *Morpho* genus.

taken, but with the fact that, by using nets with a very small mesh, fishermen are catching young immature fish before they have a chance to breed. Also there is a need for international restrictions on the sale of highly valued tropical fish for home aquariums.

On the land

Most of the major species of endangered lower animals on the land, and in its associated freshwater lakes and rivers, are reptiles. In the Americas two tortoises, the Mexican gopher tortoise *(Gopherus flavomarginatus)* and the giant Galap-

agos land tortoise *(Geochelone elephantopus)*, are at risk, and the population of the Cuban crocodile *(Crocodylus rhombifer)* has been estimated at fewer than 500 individuals. The short-necked tortoises *(Pseudemydra umbrina)* of western Australia probably total only half that number. And on the small islands off the North Island of New Zealand the unique primitive lizard-like animal called the tuatara *(Sphenodon punctatus)* — the sole remaining species of a whole order of reptiles that has remained virtually unchanged for 20 million years — is finally in danger of joining the dinosaurs.

Large land reptiles continue to suffer from the spread of man's activities and introduced predatory species. The Galapagos land tortoise *(Geochelone elephantopus,* above left) and the Javanese komodo dragon *(Varanus komodoensis)* — at 10 feet (3 meters) long the world's largest lizard — are both endangered.

Israel painted frog
Discoglossus nigriventer
Possibly extinct

Illinois mud turtle
Kinosternon flavescens spooneri
Fewer than 20

Italian spade-footed toad
Pelobates fuscus insubricus
Almost extinct

False gavial
Tomistoma schlegelii
About 100

Desert slender salamander
Batrachoseps aridus
250—500

St Croix ground lizard
Ameiva polops
About 150

Orange toad
Bufo periglenes
5,000—10,000

Watling Island ground iguana
Cyclura rileyi rileyi
200

Santa Cruz long-toed salamander
Amblystoma macrodactylum croceum
About 10,000

Indian gavial
Gavialis gangeticus
450

Endangered amphibians (far left) include various species of frogs, toads and salamanders, many of which suffer from the drainage or pollution of the waters in which they breed. Among reptiles (left), turtles and crocodiles are threatened mostly by being hunted for their shells and skins. The numbers given refer to the estimated survivors in the early 1980s.

Endangered birds

Ever since the last dodo *(Raphus* sp.) disappeared from the island of Mauritius in 1681, about 80 species of birds have become extinct. A few of them may have died out naturally, but most have succumbed as a result of human destruction, competition or other interference.

As well as these birds that can never be brought back, there are today about 210 species and subspecies in danger of extinction. The main threats are to large birds that are conspicuous and easy to kill, as well as suffering from the disadvantage of breeding slowly. Birds that are confined to islands are particularly at risk.

Direct threats

There are various kinds of direct threats. The most obvious is hunting — for food, feathers or sport. Although a significant threat to birds in many parts of the world, uncontrolled hunting is declining in northern Europe and North America, with more enlightened attitudes to conservation. But controls have come too late for some species, such as the Eskimo curlew *(Numenius borealis).* Once seen in huge migratory flocks numbering thousands of birds, it is now reduced to a handful of survivors and may well already be extinct.

The Californian condor *(Gymnogyps californianus)* is another seriously endangered American bird. One of the world's heaviest flying species, with an 8-foot (2.4-meter) wingspan, this majestic New World vulture was once scarce but widespread over much of North America; but today there are only about 30 birds in one small area of southern California. The primary reason for its decline was land cultivation (it needs huge hunting areas in which to find animal corpses),

and the small population has been further depleted by illegal shooting.

Another example of such a large, slow-breeding bird is the Steller's, or short-tailed, albatross *(Diomedia albatrus).* Japanese plume traders slaughtered more than 5 million of them in 17 years, so that it is now reduced to a small colony on Torishima Island, near Japan.

To this day, hunting in areas such as the Mediterranean accounts for many thousands of birds killed. In Italy, for instance, about 100 million songbirds such as robins and thrushes are killed every year. Such pressures can result in local extinctions or near extinctions, as with the threatened European population of the great bustard *(Otis tarda)* or the extermination of the Arabian race of the ostrich *(Struthio camelus syriacus)* in the 1960s as a result of elaborate hunting expeditions. An indirect hazard of hunting is that birds eat, with their food, the waste lead shot scattered over the ground from shotguns. It is a particular threat to swans, geese and ducks.

Rare species are threatened by unscrupulous egg collectors and taxidermists. Illegal collecting of live birds is also a problem. Exotic tropical birds suffer most; huge numbers are captured for the pet trade and for the less reputable zoos, and many birds die crammed in packing cases in aircraft. For example, about 50 of the beautiful cock-of-the-rock *(Rupicola peruviana)* die in transit from South America for every one that survives.

Another direct threat occurs when birds are killed because they are thought, often wrongly, to be a pest. The only North America parrot, the Carolina parakeet *(Conuropsis carolinensis),* was once common, but by 1904 had been hunted to extinction because of its fondness for fruit, especially the citrus fruit in the settlers' orchards (which had replaced much of the original forest

In the early 1980s, more than 20 species of birds had fallen in numbers to fewer than 50 each. Indeed, some of those illustrated here have not been sighted for several years and may by now have become extinct.

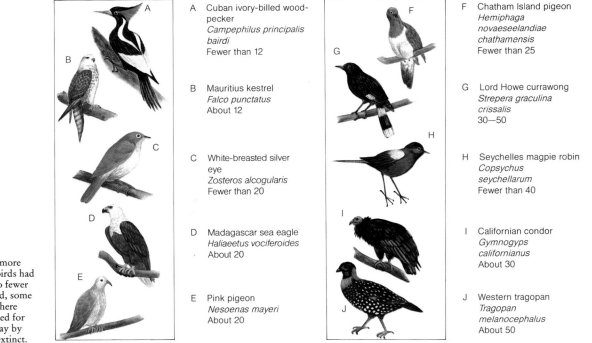

A Cuban ivory-billed woodpecker
   Campephilus principalis bairdi
   Fewer than 12

B Mauritius kestrel
   Falco punctatus
   About 12

C White-breasted silver eye
   Zosteros alcogularis
   Fewer than 20

D Madagascar sea eagle
   Haliaeetus vociferoides
   About 20

E Pink pigeon
   Nesoenas mayeri
   About 20

F Chatham Island pigeon
   Hemiphaga novaeseelandiae chathamensis
   Fewer than 25

G Lord Howe currawong
   Strepera graculina crissalis
   30—50

H Seychelles magpie robin
   Copsychus seychellarum
   Fewer than 40

I Californian condor
   Gymnogyps californianus
   About 30

J Western tragopan
   Tragopan melanocephalus
   About 50

The ne-ne, or Hawaiian goose *(Branta sandvicensis,* left), was so reduced in numbers in its island habitats that small breeding colonies have been established in wildfowl reserves. The fate of the monkey-eating Philippines eagle *(Pithecophaga jefferyi,* right) is less certain because, due to its specialized diet, it is extremely difficult to keep in captivity.

habitat). In New Zealand, the kea *(Nestor notabilis),* a parrot with a hooked, hawklike bill, has been wrongly accused of killing game birds and attacking sheep.

Indirect threats

The main damage to birds is being done indirectly by habitat destruction or alteration, or pollution. The most serious threats are to birds of tropical rainforests and wetlands. The Cuban and American race of the ivory-billed woodpecker *(Camphepilus principalis),* for example, may have reached extinction because of the destruction of the forests in which it depended on dead and dying trees for nesting. The diminutive Kirtland's warbler *(Dendroica kirtlandii)* struggled to survive in its extremely specialized habitat in central Michigan, United States, helped by careful management and strict protection. There are about 1,000 of these small yellow-breasted birds, which nest only in areas of forest that have been burned.

Wetland species such as the Japanese crane *(Grus japonensis)* and the whooping crane *(Grus americana)* are similarly threatened as drainage, pollution and acid rain from factory emissions and so on destroy their habitats. Oil pollution accounts for the deaths — often slow and miserable — of countless thousands of seabirds, such as puffins, razorbills and other auks, divers (loons) and waterfowl. Such birds are particularly vulnerable because they spend a considerable proportion of their time swimming on the surface of the sea.

Other indirect threats to seabirds include fishing nets. Danish fishing boats off Greenland are thought to have killed an estimated 500,000 of the auks called Brünnich's guillemots *(Uria lomvia)* in their salmon nets.

Pesticides have had a devastating effect on many birds, such as the peregrine falcon *(Falco peregrinus).* In the 1950s the peregrine was almost wiped out in North America and severely reduced in Europe because it ate prey that had been feeding on crops sprayed with DDT and other pesticides. Other birds that have suffered similarly include the American bald eagle *(Haliaetus leucocephalus),* the osprey *(Pandion haliaetus),* var-

ious species of pelicans, and the bald ibis *(Geronticus eremita).*

Overhead power cables kill or maim many migrating birds, especially at night. Lighthouses are also a threat to night migrants, as are some skyscrapers, and oil and gas flares can kill huge flocks at once. Cars and trucks also take a great toll of birds worldwide.

One of the worst of all indirect threats comes from animals introduced by man, especially on islands where there are no indigenous mammalian predators. Cat, rats, dogs and hogs have helped to wipe out many species. Rats, for instance, have caused the extinction of at least nine species of flightless island rails, and cats introduced in 1931 to Herokapare Island, New Zealand, have wiped out six species and reduced the total bird population from an estimated 400,000 to a few thousand. On tiny St Stephens Island, New Zealand, the entire population of the unique wren that lived there was killed in a single year (1894) by a single cat belonging to the lighthouse keeper.

A male osprey *(Pandion haliaetus)* swoops down to a female before mating. The decline of this fish-eating eagle has resulted mainly from poisoning with pesticides, although even its eggs are vulnerable to theft by unscrupulous human "collectors."

Endangered mammals

Almost every year for the past 80 years, at least one species of mammal has become extinct. Of the 4,000 or so living species of mammals, 375 are currently listed as endangered, and are given protection by international law from hunting and exploitation. Less adaptable than lower forms of life and usually less mobile than most birds, mammals are in the front line of the fight for survival.

Man has considerable commercial interest in mammals and has long been dependent on them for a variety of purposes. Whales are killed for their oil and meat, monkeys are snatched from forests for medical research, leopards and cheetahs are killed for sport and the profit from the sale of their furs. It is estimated that two million deer are killed by hunters each year in the United States. Other mammals — the giant panda and tiger, for example — are threatened by the destruction of their natural environments, caused by concentrated exploitation of valued natural resources or through land reclamation. Chemical pollution and effluent are other hazards.

Special parts of some mammals are sought by the fashion and jewelry trade. Rhinoceros horn is a highly-prized ingredient in Oriental potions; ambergris — a substance exuded by sperm whales — is used as a base for perfumes; elephant and walrus ivory is lucrative; and crocodile skin is still in demand, often on black markets. Even when animals are protected by law, poachers will go to great lengths to obtain them.

Sea mammals
The whale is a potent symbol of endangered aquatic mammals, and of the need for international management and regulation. In the last century, the herds of whales that congregated in the plankton-rich waters of the Antarctic were extensively hunted and killed. Modern intensive hunting has left many species at the edge of extinction. A hundred years ago, the blue whale — the world's largest mammal — numbered nearly 250,000; by 1965 there were a mere 2,000 left, and some experts believe that the animal is now beyond saving.

International concern over the blue whale caused Antarctic whalers to "mass hunt" smaller species. As a result, all of the eight species of whales are now in danger. The grey whale of the Pacific is near to extinction; the humpback whale has been reduced to an estimated 1,000; fin and sperm whales, as well as sei and Minke whales, are also threatened.

Protection is given to other mammals in the sea. Restrictions on hunting have saved the polar bear and various species of seals. The Arctic fox is also given limited protection.

European mammals
The fate of the European bison, the largest of Europe's mammals, was determined by a high demand for the land they roamed. By the beginning of the twentieth century, the bison was almost extinct, as were the wolves and bears that occupied the same habitats. National parks throughout Europe now offer sanctuary not only to the bison but also to the elk deer, beaver, otter and other mammals. Bears are protected in special reserves.

Once found over most of Europe, the wolf is now becoming scarce in many countries. It is rare in western Europe and extinct in 11 countries, killed for a variety of reasons. Conservation groups lobby for controls and for finance to research the wolf's environment. One scientific project in Italy resulted in the government granting complete protection to the species — there are only 100 wolves remaining there. In Spain, hunting and poisoned traps account for one in six of the rare lynx, although goverment protection is helping to stabilize the population.

African mammals
The African continent is inhabited by a spectacular range of mammals — rhinoceroses, gorillas, elephants and hippopotamuses, to name but a few. The colonization of Africa over the last two centuries sparked off a mass slaughter of many of the large mammals. During the 1880s, for example, nearly 80,000 elephants were killed for their ivory. Today the elephant is protected on many special reserves.

Many of Africa's familiar mammals are now seriously at risk. A list of threatened African mammals includes the Madagascar lemurs, okapi, gorilla, mountain zebra, eland and white rhinoceros. These animals have been given complete protection, and the giraffe, chimpanzee and some of the big cats — the cheetah most of all — are guaranteed a limited safety. There is, however, still much poaching in the large game reserves.

American mammals
Like the polar bear of northern America, the brown grizzly bear of the Rocky Mountains and

the spectacled bear of the Andes in the south are endangered mammals. Other American mammals in danger include the puma, pronghorn antelope, tapir, pamaps deer, chinchilla and vicuna. (The sale of vicuna wool has been banned in some countries to make the killing of the animals less profitable.)

National reserves have made a beginning in halting the slaughter. In a Peruvian park on the slopes of the Andes, for example, the puma, jaguar, ocelot, agouti, peccary, tapir, sloth and bush dog are among the endangered species given some protection to maintain their dwindling populations.

The United States contains some of the world's largest and best equipped natural animal reserves. On these, woodland bison, alligators, antelopes, bears and white-tailed deer are among some of the animals protected.

Australasian mammals

In Australia, the marsupials are of unique scientific interest as they are found nowhere else in the world; yet these same animals are being threatened by the spread of mechanized agriculture and the search for minerals.

Special efforts are, however, being made to preserve the koala bear and the egg-laying duck-billed platypus and echidna. Nearly 3 million square miles (8 million square kilometers) of Australian territory now lie in the boundaries of a number of excellent national parks. Some areas protect the red kangaroo. In 1965, for example, 1.5 million kangaroos were killed for their skins and meat (used for pet food). Some states do not regulate hunting, and the kangaroo is still in danger. One mammal, the striped wolflike thylacine, has already become extinct in mainland Australia and survives only in Tasmania.

Ever watchful, the Pampas deer *(Ozotoceras bezoarticus,* above left) clings to a precarious existence in the grasslands of northern South America. A white rhinoceros and its young *(Ceratotherium simum)* have to be equally alert as they break cover to go down to a water hole to drink. Hunting poses a major threat to both species.

Cuban solenodon
Atopogale cubana
Fewer than 20

Eastern cougar
Felis concolor concolor
About 20

Greater bilby
Macrotis lagotis
About 20

Northern hairy-nosed wombat
Lasiorhinus krefftii
Fewer than 40

Southern bearded saki
Chiropotes satanus satanus
Fewer than 50

Vancouver Island marmot
Marmota vancouverensis
Fewer than 100

Golden lion tamarin
Leontopithecus rosalia
Fewer than 100

Buff-headed marmoset
Callithrix flaviceps
About 100

Lower Californian pronghorn
Antilocapra americana peninsularis
About 100

Volcano rabbit
Romerolagus diazi
Fewer than 200

The most endangered mammals range in size from the Cuban species of the shrew-like solenodon to the cougar and pronghorn antelope from the eastern United States. Once the numbers of a particular species fall below about 50 — the numbers given here were estimated in the early 1980s — it is extremely doubtful whether a viable breeding population can be maintained without special protective measures being taken, and maintained, for many generations.

Conservation successes

Despite the number of animal species that have become extinct, the conservation movement has had many successes. One achievement has been the establishment of national parks and reserves. An international conference in Paris in 1968 established that such parks should be at least 1,920 acres (800 hectares) in extent, with strict prohibitions on mining, cultivation, stock-raising, hunting and fishing. There are now about 1,200 reserves in the world, covering a total of 6,700,000 acres (2,800,000 hectares). Protection is also given to animals in their natural environments.

Conservation in the sea

One success story involving a natural environment is the preservation of the polar bear. In 1955, the International Union for the Conservation of Nature recommended that all Arctic countries should curb the hunting of bears. In 1965 the Soviet Union banned the killing of polar bears and established a special reserve for them on Wrangle Island. Norway has set up a similar sanctuary. In

Canada, only license holders and indigenous Eskimos are permitted to hunt bears. Despite safari hunts outside territorial waters, the population of the animals is now increasing.

Successful treaties and special reserves have also ensured the survival of the northern fur seal, a species that lives and breeds around the islands of the Bering Sea. During the nineteenth century, the seal population was reduced to a precarious level to obtain skins for the clothing industry. Protected by law, the population of seals is currently multiplying.

On the land

Some species of deer are now ensured survival, some after a dramatic rescue. Père David's deer is named after the French missionary Armand David who, in 1865, discovered 120 of the animals living in a walled royal park in China and took some to European zoos. Successful breeding brought about a slight increase in numbers. But because of wars this initial success was mitigated

Conservation measures have succeeded in increasing the numbers of two northern mammals: the Alaskan, or northern, fur seal *(Callorhinus ursinus,* below) and the once rare saiga antelope *(Saiga tatarica,* below right) of the Asian steppes.

The scimitar-horned oryx *(Oryx dammah)* was in danger of the same fate as its Arabian cousins until, like them, it was bred in captivity. Here a pair of these graceful animals feed contentedly in the protection of a European wildlife park, far away from their native Africa.

so that by 1920 the only specimens survived at Woburn Abbey, England. Postwar breeding has swelled their numbers and Père David's deer are now in zoos throughout the world — to reduce the risk of one disease destroying them all — and more have been released in national parks.

The oryx has also been brought back from the edge of extinction. There are three different species. One is the beisa oryx, gemsbok or fringe-eared oryx. A second is the scimitar-horned oryx, whose natural habitat is on the fringes of the Sahara. The Arabian oryx is the third.

By 1960 there were only an estimated 50 Arabian oryxes from those hunted using machine guns and jeeps. All survived in the Empty Quarter of the Arabian peninsula. Three were captured in 1962 and taken to Phoenix Zoo in Arizona, where they bred successfully with other oryxes, forming the world's herd of this animal. Plans were made to return herds to the wild in the late 1980s.

Captive breeding programs are successful also in ensuring the survival of a variety of rhinoceroses. The white rhinoceros of Uganda has been saved when only 50 remained.

Many national parks in Africa offer protected sanctuary to rhinoceroses and other animals. Kruger Park in South Africa covers nearly 2,500,000 acres and contains many different habitats, ranging from humid tropical forest to near desert, hot savanna to bamboo forests. Apart from animals such as the wildebeest, Kruger Park harbors groups of other animals, such as zorils, that would otherwise be extinct.

National parks now safeguard such animals as the otter and beaver. In Asian Soviet Union, for example, the kiang (a wild ass) and the saiga (an antelope) have been rescued from extinction; indeed, so successful has been the saving of the saiga that permits can now be obtained for limited hunting. The Ussuri tiger and Himalayan bear are also preserved on Russian reserves.

Birds

European national parks and wildernesses guarantee the conservation of several varieties of eagles and owls, the peregrine falcon, and other birds such as the rose flamingo. The Camargue, a swampy wilderness in France, is one of many wetlands that has preserved a great number of the 150 species of migrating water birds. In Africa, similar sanctuaries have conserved the secretary bird, ostrich, maribou stork, pelican and many small birds that migrate the length of the continent.

There are many examples of species of endangered animals whose complete extinction has been averted yet which cannot be described as successfully conserved in the full sense. It is a testimony to the conservation movement that even this much has been achieved. From a small beginning it is hoped that such successes will eventually lead to species being taken off the endangered list and add extra momentum to the movement that preserves wildlife. And will the saving of wildlife ultimately mean the saving of humankind?

Chinese animals saved from extinction in Britain, the Père David's deer *(Elephurus davidianus)* introduced into zoos and animal parks throughout the world, were all bred from a small group that had survived in the grounds of an English country house.

Growling defiance from its solitary perch on the ice, a polar bear *(Thalarctos maritimus)* represents one of the major conservation successes.

Animal taxonomy

Kingdom	Phylum	Subphylum	Class	
Protista	Protoza	Sarcomastigophora	Phytomastigophora	Zoomastigophora
			Rhizopodea	Actinopodea
		Ciliophora	Kinetofragminophora	Oligohymenophora
			Polyhymenophora	
		Sporozoa	Sporozea	Piroplasmea
		Cnidospora	Myosporidea	Microsporidea
Metazoa	Porifera		Calcaria	Hexactinellida
			Demospongiae	Sclerospongiae
	Cnidaria		Hydrozoa	Scyphozoa
			Anthozoa	
	Ctenophora		Tentaculata	Nuda
	Platyhelminthes		Turbellaria	Trematoda
			Cestoda	
	Entoprocta			
	Rhyncocoela		Anopla	Enopla
	Gnathostomulida			
	Gastrotricha			
	Rotifera		Digononta	Monogonta
	Echinorhyncha			
	Nematoda		Aphasmida	Phasmida
	Nematomorpha		Gordioidea	Nectonematoidea
	Acanthocephala			
	Annelida		Polychaeta	Oligochaeta
			Hirudinea	
	Echinodermata		Asteroidea	Ophiuroidea
			Echinoidea	Holothuroidea
			Crinoidea	
	Mollusca		Gastropoda	Bivalvia
			Cephalopoda	Monoplacophora
			Polyplacophora	Aplacophora
			Scaphopoda	
	Arthropoda	Chelicerata	Merostomata	Pycnogonida
			* **Arachnida**	
		Crustacea	Cephalocarida	Branchiopoda
			Ostracoda	Mystacocarida
			Copepoda	Branchiura
			Cirripedia	* **Malacostraca**
		Uniramia	* **Insecta**	Chilopoda
			Diplopoda	Pauropoda
	Tardigrada			
	Linguatulida			
	Echiura			
	Bryozoa			
	Priapulida			
	Phoronida			
	Brachiopoda			
	Sipunculidea			
	Pogonophora			
	Chaetognatha			
	Chordata	Hemichordata	Pterobranchia	Enteropneusta
		Urochordata		
		Cephalochordata		
		Vertebrata	Agnatha	Elasmobranchiomorphi
			* **Osteichthyes**	Amphibia
			Reptilia	* **Aves**
			* **Mammalia**	

* Largest Animal Classes

Class (or **Subclass**)	Superorder (or **Infraclass**)	Order		
Arachnida		Scorpiones	Pseudoscorpiones	Solifugae
		Palpigradi	Uropygi	Schizomida
		Amblypygi	Araneae	Ricinuclei
		Opiliones	Acarina	
Malacostraca (**Phyllocarida**) (**Eumalacostraca**)		Leptostraca		
	Syncarida	Anaspidacea	Bathynellacea	
	Hoplocarida	Stomatopoda		
	Eucarida	Euphausiacea	Decapoda	
	Peracarida	Mysidacea	Cumacea	Tanaidacea
		Thermosbaenacea	Spelaeogriphacea	Isopoda
		Amphipoda		
Insecta (**Apterygota**) (**Pterygota**)		Protura	Thysanura	Collembola
		Ephemoroptera	Odonata	Orthoptera
		Isoptera	Plecoptera	Dermaptera
		Embioptera	Psocoptera	Zoraptera
		Mallophaga	Anoplura	Thysanoptera
		Hemiptera	Homoptera	Neuroptera
		Coleoptera	Strepsiptera	Mecoptera
		Trichoptera	Lepidoptera	Diptera
		Hymenoptera	Siphonoptera	
Osteichthyes (**Actinopterygii**)	Chondrostei	Acipenseriformes	Polypteriformes	
	Holostei	Lepisosteiformes	Amiiformes	
	Teleostei	Elopiformes	Anguilliformes	Notacanthiformes
		Clupeiformes	Osteoglossiformes	Mormyriformes
		Salmoniformes	Cetomimiformes	Gornorynchiformes
		Ctenothrissiformes	Cypriniformes	Siluriformes
		Percopsiformes	Batrachoidiformes	Gobiesociformes
		Lophiiformes	Gadiformes	Atheriniformes
		Beryciformes	Zeiformes	Lampridiformes
		Gasterosteiformes	Channiformes	Synbranchiformes
		Scorpaeniformes	Dactylopteriformes	Pegasiformes
		Perciformes	Mastocembeliformes	Pleuronectiformes
		Tetradontiformes		
(**Sarcopterygii**)		Crossopterygii	Dipnoi	
Aves (**Neornithes**)	Neognathae	Struthioniformes	Casuariiformes	Apterygiformes
		Rheiformes	Tinamiformes	Sphenisciformes
		Gaviiformes	Podicipediformes	Procellariformes
		Pelecaniformes	Ciconiiformes	Anseriformes
		Falconiformes	Galliformes	Gruiformes
		Charadriiformes	Columbiformes	Psittaciformes
		Cuculiformes	Strigiformes	Caprimulgiformes
		Apodiformes	Coliiformes	Trogoniformes
		Coraciiformes	Piciformes	Passeriformes
Mammalia (**Prototheria**) (**Theria**)		Monotremata		
		Marsupalia		
	(**Metatheria**) (**Eutheria**)	Insectivora Edentata	Pholidota	Chiroptera
		Primates Rodentia	Lagomorpha	Cetacea
		Carnivora Tubulidentata	Hyracoidea	Proboscidea
		Sirenia Perissodactyla	Artiodactyla	

145

GLOSSARY

In the following glossary, small capital letters (e.g. NUCLEUS) indicate terms that have their own entries in the glossary.

A

allantois A membrane in the EMBRYO of TETRAPODS. It carries a large number of blood vessels which connect the blood system of the mother's PLACENTA with the embryo and, in reptiles and birds, allows gas exchange between the SHELL and the embryo.

allele A GENE with a pair member, both occupying the same loci on homologous CHROMOSOMES. They pair during MEIOSIS and can MUTATE one to the other.

alternation of generations Found in lower plants and some animals, especially coelenterates, it involves organisms of distinctly different body form, such as POLYPS and MEDUSAE, each of which gives rise to the other.

alula Also known as a "bastard wing," this small bunch of feathers is attached to the thumb-bone of a bird's wing. At low flight speeds it can be raised to form a "slot" which controls air flow over the upper surface of the wing, reducing the stalling speed.

ametabolous Describes those insects which hatch from eggs as miniature replicas of their wingless parents and simply grow in size between each molt, without METAMORPHOSIS. The process occurs in the subclass Apterygota, a group of primitive wingless insects, such as silver fish.

amino acid A member of a group of organic acids containing an amino group, and a component of DNA molecules in the CELL NUCLEUS. About 25 are known to be found in PROTEINS of which they are the building blocks, combining in different orders to form the proteins.

amnion A fluid-filled sac and the innermost membrane in which the EMBRYOS of TETRAPODS develop.

asexual reproduction The formation of a new individual from a single parent. This may be by PARTHENOGENESIS, by budding or by fragmentation.

B

bacterium A microscopic uni- or multicellular organism. Bacteria vary in shape (the basis of their classification) and motility. They are generally considered to be plant-like although they lack CHLOROPHYLL. They feed on plant and animal tissues (as well as inorganic matter), decomposing

it and thus releasing in water and soil the nutrients which support larger and more complex organisms.

baleen Plates of horny material hanging from the palate of toothless whales. The inner edge is fringed and forms a filter. Seawater drawn into the open mouth is squirted out through the baleen, food contained in the water is kept behind.

bastard wing See ALULA.

benthic Describes organisms that inhabit the bed of a lake or the sea.

bilateral symmetry The symmetry of most active animals with an elongate body. At one end is a head in which the main sense organs and BRAIN are assembled; behind it the body carries paired organs such as limbs, which are mirror images of each other.

bilharzia Also known as schistosomiasis, this parasitic disease, widespread in the tropics, is carried by flukes of the genus *Schistosoma*. It is transmitted in water when the LARVA of the fluke may burrow through the skin. Like most parasitic diseases, it is debilitating and may be fatal.

binary fission Reproduction of a CELL by splitting into two equal parts, common in many single-celled organisms.

binomial classification The system of classifying plants and animals by which each SPECIES is given a unique combination of two names: the genus, or group name, and the species name. The genus name always begins with a capital letter, the species with a small letter; e.g. *Panthera leo*, the lion.

blubber Fat lying beneath the skin of seals and whales. Consisting of oil-filled cells, it is a poor conductor of heat and helps to preserve body warmth. It also acts as a food reserve.

bone Vertebrate skeletal material, formed largely from COLLAGEN, phosphates and calcium salts. It is hard, but flexible to some extent.

brain The forepart of the main nervous system in BILATERALLY SYMMETRICAL animals. It is housed in the head, close to the principal sensory organs, such as the EYES, ears and nose. To a greater or lesser extent the brain coordinates the activities of the entire body.

C

carnassial In carnivores, the first lower molar tooth and the last upper premolar teeth. Together they form large shearing blades and are used for slicing flesh.

cartilage A skeletal tissue in vertebrates containing COLLAGEN fibers in a matrix composed largely of carbohydrates. It is softer and more flexible than BONE.

cell The basic unit of organic tissues. Cells are bounded by a cell wall or membrane and contain a NUCLEUS, which is derived by division from a previous cell nucleus and holds the CHROMOSOMES, and CYTOPLASM. Some organisms are unicellular but most comprise many specialized cells.

cellulose A complex carbohydrate material, with a fibrous structure that makes the cell walls of plants rigid.

cephalothorax The fused head and forebody found in arachnids and some crustaceans.

chiasmata The crossing-over or exchange of genetic material between CHROMOSOMES during MEIOSIS. It can lead to the production of new varieties within a SPECIES.

chitin The tough, fibrous EXOSKELETON of arthropods. It is relatively inelastic and therefore dictates the jointed limb and body structure of arthropods and their system of growth by molting.

chlorophyll A green pigment found in most plants, contained in the CHLOROPLASTS. It absorbs light from the Sun and by photosynthesis converts it into chemical energy used to build up carbohydrates (sugars and starches).

chloroplasts Small, pigment-containing bodies found in plant CELLS. They contain CHLOROPHYLL as well as other pigments which give plants their color.

choanocyte A collar CELL found in some sponges, in which a flagellum arises from a PROTOPLASMIC collar. The beating of the flagella moves water containing food and oxygen through the sponge's body.

chromatid One of two long filaments of GENETIC material on a duplicated CHROMOSOME, visible during MITOSIS and MEIOSIS.

chromatin The nucleoprotein material of which CHROMOSOMES are made.

chromatophore A pigment-containing cell found in many animals, such as squids, chameleons and flatfish. The pigment level can be altered rapidly, which enables the animal to change color.

chromosome Found in the NUCLEUS of every plant and animal CELL, chromosomes are composed of CHROMATIN, and carry the GENES which are the units of inheritance. The number of chromosomes varies in different SPECIES, but in all, the sex cells (GAMETES) contain half of the normal complement. On fertilization the chromosomes from the male and female gametes come together to give the new individual its complete inheritance pattern, received in part from each parent.

clavicle The collarBONE found in vertebrates. The clavicles have been lost in many mammals, but are found in birds as a single, fused element called the furcula, or wishbone.

cleavage The subdivision of a fertilized ZYGOTE during its development to form a hollow sphere of small cells called the blastula.

clitellum A saddle-like structure found in some sexually mature annelids. It produces a mucous sheath round mating animals and binds the cocoon round the fertilized eggs.

cloaca The end of the gut in vertebrates (except for non-marsupial mammals) into which the digestive and reproductive tracts open.

cnidocil The trigger hair of a NEMATOCYST, or stinging cell.

coelenteron The body cavity of coelenterates. It has one opening only and through it food enters and waste products and sperm or eggs are ejected.

coelom The main body cavity of many complex animals.

collagen A PROTEIN which forms intercellular fibers in the skeletal tissues of vertebrates. Collagen fibers have great tensile strength, but little elasticity.

commensalism The association of two different organisms where mutual advantage may result, although the partners may also be capable of independent existence.

compound eye Found in crustaceans and insects, compound eyes consist of a number of separate lens systems (more than 10,000 in each eye in some insects), each of which accepts information from a very small part of the surroundings. They may be color-sensitive, but lack visual acuity compared with cephalopod or vertebrate EYES.

conjugation The partial fusing of two individual ciliate protozoan cells, when nuclear material is exchanged, during the process of SEXUAL REPRODUCTION.

coracoid This strutting bone in the PECTORAL GIRDLES of primitive land vertebrates lies between the sternum and the outer ends of the CLAVICLES. They have been totally lost in mammals, but are present in birds.

cornea The clear, tough, protective layer of connective tissue that covers the lens of the EYE in vertebrates.

corona The ciliated "crown" of rotifers. The activity of the CILIA is the means of locomotion and food collection in these creatures.

crop A thin-walled, distendable sac in the esophagus of many birds, which functions as a food store.

crystalline style A rod of PROTEIN containing a starch-digesting ENZYME found in the gut of some mollusks. The free end projects into the stomach and is worn away as the enzyme is mixed with the gut contents.

cuticle The protective surface layer that covers plants and animals.

cypris The fully developed LARVA of a barnacle, when it is ready to settle on a surface.

cytoplasm The watery substance that surrounds the NUCLEUS of a CELL and which contains bodies such as mitochondria, and Golgi bodies. Together with the nucleus they form the PROTOPLASM of the cell.

D

delayed implantation A pause in the development of the EMBRYO of some mammals which, after early CELL division, remains unattached to the wall of the UTERUS for a period of up to several months. It probably allows courtship and mating, as well as birth of the young, to occur at times when food supplies are abundant.

deoxyribonucleic acid See DNA.

dewclaws The toes of a DIGITIGRADE or UNGULIGRADE animal that do not normally touch the ground. In carnivores, such as dogs, the first toe of the forelimb forms a dewclaw; in many deer, the second and fourth digits of all feet are dewclaws.

diaphragm A sheet of tissue which separates the chest cavity from the abdomen, found in mammals. It can be arched or flattened to improve the efficiency of breathing.

digitigrade The method of TETRAPOD locomotion in which an animal walks on its toes, thus increasing its stride by the length of the hand or foot bones. Dogs and cats are digitigrade.

dioecious A term applied to plants and animals which have separate sexes.

displacement activity An inappropriate form of behavior performed as a result of conflicting drives. It usually appears as an exaggerated form of normal activity.

DNA (deoxyribonucleic acid) A compound found in plant and animal CELLS, formed from nitrogenous bases arranged in a double helical form and capable of self-replication. It carries GENETIC information in the arrangement of these bases. Each sequence of three bases forms the code for one AMINO ACID.

E

ecdysis The shedding of the EXOSKELETON in arthropods, necessary in this group to allow growth. Under HORMONAL control, the body covering is split off, allowing a new one, already produced beneath the old, to expand and then harden.

echo-location A method of short-distance navigation and sometimes food-finding, used by some nocturnal and aquatic animals. Short bursts of sound are produced by the animal which, if they encounter an obstruction, return an echo which is detected by the animal. Bats and whales echo-locate using ULTRASONIC sound.

eclipse period A dull-colored, post-mating plumage phase in many birds. In ducks and geese this phase is accompanied by a simultaneous molt of the flight feathers, so that they are temporarily flightless.

elytron The thickened and stiffened forewing in beetles.

embryo The pre-birth or pre-hatching stage of a developing animal.

endoskeleton The body support that lies within the muscular structure. It is best developed in vertebrates as a SKELETON, but is also seen in some other groups.

enzymes Proteinaceous substances that are produced by living CELLS, vital to the processes of METABOLISM.

epidermis The outermost layer of CELLS of plants and animals. In vertebrates it is several cells thick and in land-living forms the outer layers are often dead and horny. In invertebrates it is one cell thick and often secretes a CUTICULAR protection.

estivation Dormancy during hot dry seasons.

estrus The period of greatest sexual receptiveness in a female mammal, normally coinciding with ovulation.

exoskeleton The bodily support that lies outside the muscular structure. It is best developed in arthropods and mollusks.

eye An organ for light-reception, varying from a simple cluster of light-sensitive CELLS, such as the OCELLUS or the COMPOUND EYE, to a complex system with one or more light-gathering lenses focusing on special cells sensitive to light intensity and color.

F

fetus A mammalian EMBRYO in the later stages of its development.

filter-feeding A method of feeding used by many marine invertebrates in which minute organisms are removed from large quantities of water that is strained through GILLS or some other sifting mechanism.

food vacuole A fluid-filled space found in single-celled animals, created when food is engulfed. The fluid contains ENZYMES which digest the food. The waste flows out from the vacuole into the CELL CYTOPLASM.

frenal hooks Tiny hooks along the edges between the two wings of hymenopterans (bees and wasps, etc.) which hold the two wings on each side of the body together so that they beat as a single unit.

furcula See CLAVICLE.

G

gamete A reproductive CELL with half the number of CHROMOSOMES normal for the SPECIES (haploid). It is either a mobile male cell (sperm) with reduced CYTOPLASM, or an immobile female cell (ovum) with hugely increased cytoplasm. They fuse during fertilization when a diploid cell (ZYGOTE), with a full complement of chromosomes, is formed.

ganglion A small knot of nerve tissue from which nerve cords arise. In invertebrates it may form the BRAIN. In vertebrates ganglia occur in the peripheral nervous system and as some of the nerve connections of the brain.

gastrulation The movement of CELLS in the early stages of the development of an EMBRYO to the positions in which they will form the internal organs of the animal.

gemmule A bundle of CELLS which are capable of developing without fertilization into a new organism. They may be a means for an individual, the major part of whose body dies in adverse circumstances, to survive until the environment is suitable for further activity.

gene The basic unit of heredity, formed from a sequence of bases on a DNA chain. Each gene has a definite position of the CHROMOSOME and may occur as an ALLELE.

genotype The GENETIC (hereditary) constitution of an individual.

genus See BINOMIAL CLASSIFICATION.

gestation The period of development between conception and birth in mammals.

gill The RESPIRATORY organ of aquatic animals. Gills are usually projections from the body wall or gut, and are often complex in shape, so that they offer the maximum surface area for gas exchange between the blood which flows through them and the water which flows round them.

gill-book Found in some aquatic arachnids, these plates of tissue resemble the pages of a book; they are set in a cavity where they can be bathed in water from which they extract oxygen and in which they release carbon dioxide.

gizzard The muscular stomach of birds and some other animals, which breaks up food prior to digestion, sometimes aided by swallowed stones.

grooming This care of the body surface, usually by scratching or licking, may be performed by a pair of animals on each other when it may be an important force in pair bonding or maintaining social cohesion within a group.

guard hairs The thick hairs which overlie the fine underfur in many mammals. In cold or wet conditions air is trapped close to the skin, and the guard hairs (which are often greasy and waterproof) clamp down and prevent the escape of this warm blanket.

H

halteres The hind wings of flies (Diptera) which have been reduced to tiny, club-shaped structures, used in flight as gyroscopic balancing devices.

heart The strong, muscular pump that drives blood through animals. In mammals and birds it has a complex of chambers which separate out deoxygenated blood and send it to the lungs from which it returns oxygenated to supply the body with its needs. In lower vertebrates the oxygenated and deoxygenated blood is mixed and in invertebrates it plays little or no part in the distribution of oxygen throughout the body.

hemimetabolous Describes insects with an incomplete METAMORPHOSIS. When they hatch, they resemble the parents, although they do not develop functional wings until mature. They grow by a series of molts.

hemocoele A closed body cavity containing blood. It is well-developed in arthropods and mollusks. Unlike the true COELOM, it never contains reproductive cells.

hemocyanin A RESPIRATORY pigment containing copper, found in the blood of most arthropods and mollusks. It gives the blood of these animals its green color.

hemoglobin An iron-containing RESPIRATORY pigment found in the blood of most vertebrates and some invertebrates.

hermaphrodite An animal possessing both male and female sexual organs. These organs are usually functional at different times, preventing self-fertilization.

hibernation A state of torpor which POIKILOTHERMAL animals enter when the surrounding temperature drops below a certain critical point. A few mammal and bird SPECIES also hibernate, but they do so for long periods (up to 7 months in some rodents). Their temperature, heart-beat and breathing rates fall dramatically and they use very little energy in staying alive.

holometabolous Describes insects which undergo a complete METAMORPHOSIS from an egg to a LARVA, a PUPA and finally an adult.

homeothermal Describes animals that are "warm-blooded," with an internal thermostat that keeps their body temperature within very narrow limits, whatever the temperature of their surroundings.

hormones Substances produced by glands, in animals, and released into the bloodstream where they have an important role in body functions, such as growth, reproduction and digestion.

host An individual on which a PARASITE feeds.

hybrid The result of successful cross-fertilization between members of different SPECIES. Hybrids among animals are rare in nature and are almost always sterile.

hydrostatic skeleton The body support found in some invertebrates, which depends on the movement of liquid inside the cells to maintain body shape or to change position.

hyoid A system of small bones in the base of the tongue in TETRAPODS, derived from a GILL arch in fish ancestors. In some SPECIES they have become enlarged to support a long insect-catching tongue, as in woodpeckers, or a resonating voice box, as in howler monkeys.

hypothalamus A part of the BRAIN of vertebrates, which is thought to contain, among other things, the mechanism of body temperature control.

I

iliac crest The expanded upper part of the ilium, which is part of the PELVIC GIRDLE of TETRAPODS. It is particularly important in humans as the area of attachment of the muscles which allow them to stand and walk upright.

imprinting The process by which many animals learn to recognize their own kind. It occurs at an early stage of development as a response to a narrow range of stimuli to which the young individual is normally exposed.

incubation The period between the laying and hatching of eggs, when they are kept warm, either by the Sun, rotting vegetation, or the warmth of the parent's body.

instinct Non-learned or innate behavior patterns, often very complex, usually triggered by specific stimuli.

K

keratin A tough, fibrous PROTEIN material found in TETRAPODS. It forms the outer layer of the skin, and hair, hoofs and horny structures in mammals, feathers in birds, and SCALES in reptiles.

L

larva The young stage of many OVIPAROUS animals which have a distinctly different body form from the adult, usually related to a specialized way of life. For example, in insects the larvae may be the feeding stage, in many crustaceans, the dispersal phase. At the end of the larval life, they rapidly METAMORPHOSE to the adult state.

lateral-line system A series of sense organs on the flanks and sides of the head in agnatha, fishes and amphibians. Each nerve ending, housed in a bony pit, is sensitive to pressure changes, so that the animals can detect, through water movement, the presence of potential enemies or prey.

lung The RESPIRATORY organ of land-living vertebrates and some mollusks. Air is drawn into a cavity, the skin of which is well-supplied with blood vessels, so that gas exchange can take place there.

lung-books The breathing organs of some arachnids. They consist of a series of thin flaps of tissue, like the pages of a book, projecting into a cavity in the body wall. Blood flows through them and gas exchange takes place at their surface.

M

macronucleus In Ciliophora, a large NUCLEUS which divides MEIOTICALLY and disappears during CONJUGATION.

Malpighian tubules Excretory glands found in the hind-gut of insects, arachnids and some other arthropods.

mammary glands MILK-producing glands of female mammals. Their state usually varies with the ESTRUS cycle and their growth and milk production is controlled by various HORMONES.

mandible The lower jaw of vertebrates, and in some invertebrates (such as insects) the major pair of crushing mouthparts.

maxilla In vertebrates, this forms part of the upper jaw; in invertebrates, it forms one of the pairs of mouthparts that lies behind the MANDIBLES.

medusa A free-swimming, jellyfish-like coelenterate which reproduces sexually. In many cases, however, the eggs of medusae give rise to POLYPS which form further medusae by ASEXUAL division.

meiosis Two stages of CELL division that result in the formation of GAMETES. (See also MITOSIS.)

mesogloea A jelly-like substance found between the ectoderm and endoderm (the outer and inner tissue layers) of coelenterates.

metabolism Life processes which involve both the breakdown (catabolism) of organic compounds to liberate energy for various activities, and also the build-up (anabolism) from simple materials of the complex tissues of an organism.

metamorphosis The change which occurs between the immature, LARVAL stage of an animal's life and its mature, reproductive phase. Complete metamorphosis involves a major change in form, and takes place within the shelter of a PUPAL case. Incomplete metamorphosis is more gradual until wing growth and sexual development are accomplished.

micronucleus A small nucleus found in Ciliophora which divides MITOTICALLY and provides the nuclear material for CONJUGATION. After conjugation the MACRONUCLEUS is reformed from the ZYGOTIC (micronuclear) material.

milk Produced in the MAMMARY GLANDS of female mammals, milk contains fats, sugars and PROTEINS and is used to feed the young.

mitosis Normal CELL division in which CHROMOSOME material is doubled, so that daughter cells carry the same GENETIC information as the parent cells.

mutation A spontaneous, irreversible GENETIC change. It normally occurs during CELL division and can affect any cells of the body. Most mutations are harmful and kill the organism, but those that are not lead to variation in the SPECIES.

myoglobin An oxygen-carrying pigment, related to HEMOGLOBIN, contained in the muscles of mammals, especially those such as seals and whales, which hold their breath for long periods.

N

nectar guides Brightly colored lines or patches on petals which lead insects towards the nectar which usually lies at the center of the flower. In their search the insects brush against pollen which they carry to other flowers to effect pollination.

nematocyst The stinging cell of a coelenterate. Thickly scattered on the tentacles, nematocysts consist of a poison cell in which a long, barbed hollow tube lies coiled, and above which lies a CNIDOCIL, or hair. When this hair is triggered, the hollow tube shoots out and cuts the skin of the prey, allowing the poison to affect the nervous system.

nidifugous Describes young birds which leave the parental nest shortly after hatching, such as ducks and game birds.

notochord An internal rod, made of a stiff, jelly-like substance, found in the LARVAE or EMBRYOS of vertebrates. It lies beneath the nerve cord and above the gut and supports the muscle blocks of the body in all chordates.

nucleus The controlling center of all cellular activity, containing the CHROMOSOMES, the nucleolus and the nucleoplasm.

O

ocellus A simple EYE — a cluster of photoreceptors — with a single lens system, found in arthropods and some other invertebrates.

ommatidium A single element of the COMPOUND EYE of insects and crustaceans.

operculum A cover used in many animal structures, such as the lid of a NEMATOCYST; the horny cover with which some gastropods close their shell opening against desiccation or enemies; and, in bony fishes, the large bony plate which protects the GILLS.

ovary The egg-forming structure in a female or HERMAPHRODITE organism.

oviduct The tube connecting the OVARY to the UTERUS in female mammals, and carrying eggs out to the body in other vertebrates.

oviparous Describes animals that lay eggs which hatch once they are laid.

ovoviviparous Describes animals in which the EMBRYO develops in an egg membrane, until it hatches inside the mother.

P

parasite An organism which lives on another at its expense. Ectoparasites, such as fleas, live on the surface of their HOST; endoparasites live internally, and include tapeworms and hookworms.

parthenogenesis The development of an unfertilized ovum into a new individual. In some animals this is the usual method of reproduction, but the pattern is often broken, by a sexual generation which allows a recombination of GENETIC material within the population.

pectoral girdle The BONES supporting the forelimbs in TETRAPODS, consisting of shoulder blades (scapulas), collar bones (CLAVICLES) and, in some cases, CORACOIDS.

pelvic girdle The hip girdle of TETRAPODS composed, in most cases, of the ischium, the pubis and the ilium. It supports the abdomen, the base of the tail, and the upper part of the leg.

pentadactyl Meaning "five fingered," this term refers to the limbs of early land vertebrates, which had five digits on each. This number has been reduced in many modern vertebrates, but the basic pattern is the same, so the term is still used.

peristalsis Waves of muscular contraction moving along tubular organs, moving and mixing the contents.

phenotype The physical form of an organism.

pheromone A chemical, usually a scent, secreted by an organism which stimulates a particular response in another individual. Pheromones play a large part in the courtship and mating behavior of many animals.

photosynthesis The process by which green plants transform the light energy from the Sun into chemical energy and use it to build carbohydrates. It occurs in the CHLOROPLASTS.

phytoplankton The plant component of the PLANKTON. These unicellular organisms are the basic food for all ocean-dwelling animals and, through their PHOTOSYNTHETIC activity, they provide much of the free oxygen in the atmosphere.

placenta A temporary organ which develops in the UTERUS of most pregnant mammals through which the EMBRYOS are nourished, receive oxygen and eliminate waste.

plankton The term covers all floating aquatic organisms. Many planktonic animals are LARVAL forms, but some are permanent members of the plankton. Most are microscopic, but a few, such as jellyfish, are large.

plantigrade The style of locomotion in TETRAPODS where the whole length of the foot is placed on the ground, as in man, most insectivores, and some reptiles.

pneumostome The RESPIRATORY pore of terrestrial mollusks, which is found behind the head, on the right side of the body.

poikilothermal Describing animals that are "coldblooded." Their body temperature is not low, but matches that of their surroundings, fluctuating with the time of day or year.

polyembryony The formation of more than one EMBRYO from a single ZYGOTE by fission at an early stage.

polyp The fixed stage in the life of many coelenterates. In some, such as sea anemones and corals, it is the only adult form; but, in others, such as hydrozoa, it is the ASEXUAL reproductive phase.

proglottid A single reproductive segment of a tapeworm. In its early life it is connected to a chain of similar segments. When mature and full of ripe eggs, it breaks away and leaves the HOST's body with the feces.

prognathous Projecting jaws, as found in baboons, apes and some early humans.

protein A complex organic compound, formed from large numbers of AMINO ACIDS. Proteins play an important part in the formation, maintenance and regeneration of tissues.

protoplasm The contents of a cell containing the CYTOPLASM and the NUCLEUS.

pseudopodium A temporary extension of a CELL caused by the directional flow of PROTOPLASM within it. It is the method of locomotion and feeding in ameboid cells.

pupa The stage between the LARVAL feeding form and the adult reproductive form in HOLOMETABOLOUS insects. During the pupal stage, the body tissues are almost completely broken down and reconstructed round a group of special CELLS.

R

rachis The main part of the shaft of a feather, which carries the vane made of barbs and barbules.

radial symmetry A circular body form in which there is no "head end" to the body and no concentration of BRAIN or nervous tissue. It is usually found among inactive animals.

radula The "tongue" of mollusks. It carries large numbers of horny teeth, which are used to rasp food.

reflex An involuntary, unlearned, stereotypic response to a stimulus.

respiration The method by which an organism oxygenates its tissues and rids itself of unwanted carbon dioxide. In invertebrates, gases may diffuse through the whole of the outer surface or via TRACHEAE to the tissues. In larger animals specialized organs have evolved, including GILLS for aquatic animals and LUNGS for terrestrial species.

retina The light-sensitive layer at the back of the EYE of vertebrates and cephalopods. It includes two types of cells — rods, which are sensitive to dim light, and cones, which are concerned with the perception of colors.

ribonucleic acid See RNA.

RNA (ribonucleic acid) The material in a CELL, composed of a single nucleotide chain, that organizes and governs PROTEIN synthesis.

royal jelly A highly nutritious secretion from the pharyngeal glands of young worker honey bees. It is fed to all the LARVAE in the colony for the first few days of their lives, after which most of them are switched to a pollen diet. Larvae which are to become queens continue to be fed solely on royal jelly.

rumination The method of digestion used by cud-chewing ungulates. They swallow their food unchewed, which goes into the rumen where it is partly broken down by BACTERIA. Later it is returned to the mouth to be masticated completely and when swallowed it by-passes the rumen.

S

scales In bony fishes, scales are fine slips of bone formed in the dermal layer of the skin and overlapping like tiles; they may be smooth-edged (cycloid), serrated-edged (ctenoid), or bony (ganoid). Scales in reptiles are made of KERATIN and on the wings of insects they are composed of CHITIN.

scolex The head of a tapeworm which is armed with hooks, by which it attaches itself to the wall of the HOST's intestine. The PROGLOTTIDS grow from the scolex.

sedentary Describes animals which remain in the same living area, such as nonmigratory birds which, even so, move through several acres of territory. Among the invertebrates it describes animals such as oysters or barnacles which are fixed to one point of the substrate.

segmentation Repetition of a structural pattern along the length of an animal's body or appendage, seen most clearly in annelids and arthropods. Each unit is referred to as a segment.

sesamoid A BONE developed within a tendon, especially in mammals, such as the kneecap, and the "thumb" bone of the giant panda.

sessile Describes animals that are attached to the substrate, as are barnacles or oysters.

sexual dimorphism The differences between the male and female form of SPECIES. Often the differences are slight but sometimes they are so great that the male and female have been thought to be of unrelated species.

sexual reproduction The formation of a new individual from a fertilized ovum, or SYNKARYON. At conception, a male GAMETE fuses with a female gamete. Because during MEIOSIS there is considerable shuffling of GENETIC material, each gamete differs from all others. The individual formed as a result of sexual reproduction is genetically unique and even the offspring of the same parents differ from each other.

shell The protective armor of many animals, including some vertebrates such as tortoises, or the outer covering of the egg of a reptile or bird. The term mainly refers to the strong EXOSKELETON of invertebrates, particularly crustaceans, mollusks and brachiopods.

shoaling In fishes, the habit of forming large, tightly-knit groups composed of individuals of the same age and size. The purpose may be safety — predators may think the group to be one large animal, too big to attack safely, and if an attack is made the large number of individuals may bewilder the predator.

skeleton The BONY internal support of vertebrates. It consists of the vertebral column which supports the head at the anterior end, a PECTORAL GIRDLE to which the forelimbs are attached and a PELVIC GIRDLE which carries the hind limbs. Fishes have fins instead of girdles.

sonar See ECHO-LOCATION.

species The basic unit of animal classification consisting of populations of organisms sufficiently similar to permit free interbreeding in the wild. (See also BINOMIAL CLASSIFICATION).

spermatophore A packet of sperm produced by the males of some species in which there is internal fertilization but in general no intromittant organ. In many cases complex mating behavior patterns have evolved to ensure that the female is correctly positioned to pick up the spermatophore in her CLOACA.

spicules The skeletal support of sponges. They may be of a horny material as in bath sponges, silicaceous in glass sponges, or chalky as in calcareous sponges.

spinnerets   Small raised organs at the hind end of a caterpillar or spider, through which silk is produced. Each one contains minute outlets through which the silk is forced in liquid form, in chains of molecules. The silk polymerizes on contact with the air.

spiracle   In cartilaginous fishes, this small opening lies in front of the GILLS. In bony fishes, it is dorsally situated and is used to take water into the gill chambers. It is also a breathing pore found on the sides of insects which opens into the TRACHEAL system.

spore   A dormant form which many protozoans adopt allowing them to withstand long periods of extremes of heat or cold which would be lethal at other times. They return to a normal, active form when conditions are appropriate.

stapes   One of the three small BONES of the mammalian inner ear. Reptiles, amphibians and fishes have a comparable bone, but are lacking the other two.

statocyst   A more or less spherical organ of balance which contains a granule of hard material which stimulates sensory CELLS as the animal moves.

sternum   The breastbone in land vertebrates, to which the ribs, CLAVICLES and CORACOIDS (if any) are attached.

suspension feeder   An animal which feeds on minute food particles strained from the water in which it lives.

swim-bladder   A gas-filled sac in the abdominal cavity of bony fishes which enables them to achieve neutral buoyancy. Its presence means that the animal need not spend energy in maintaining its level in the water, but rather can use all of its locomotive power in moving forward.

symbiosis   An association of two or more different organisms which are mutually interdependent and cannot live alone.

synkaryon   The diploid ZYGOTE NUCLEUS, formed by the fusion of GAMETIC nuclei.

synovial capsule   A bag of connective tissue surrounding any free-moving joint such as a shoulder, elbow or wrist. It contains synovial fluid which lubricates the CARTILAGE coverings of the ends of the BONES.

T

tapetum   A reflecting layer at the back of the EYE of some vertebrates, particularly those which are nocturnal. It enables the eye to make the fullest possible use of any available light.

taxon   A classificatory unit, such as a SPECIES, GENUS or family.

tergum   The thickened CUTICLE on the upper surface of a segment of an insect or crustacean.

testis   The sperm-forming structure in a male or HERMAPHRODITE organism.

tetrapod   Literally meaning "four feet," this term is used to denote any land vertebrate.

torsion   The twisting of the visceral mass through 180° which takes place in the LARVAL stage of gastropods, bringing the anus from its original posterior position into an anterior position above the head. The advantage appears to be that the mantle cavity then provides space into which the animal can retract its soft parts, within the protection of the SHELL, closed in many cases by the horny OPERCULUM.

trachea   A breathing passage in insects and some arachnids. It conducts air from the SPIRACLES on the skin into the body, where it branches to form fine tubes called tracheoles, where gas exchange takes place. In higher vertebrates, the trachea is the windpipe.

trochophore   A planktonic LARVA of polychaetes and some mollusks. Roughly spherical in shape, it has a ring of cilia round the body, the beating of which stabilizes the animal in the water and brings food to its mouth.

U

ultrasonic   Describes high-frequency sound waves above the hearing of human ears, used by bats and whales in ECHO-LOCATION.

unguligrade   A method of mammalian locomotion in which the animal walks on the tips of its toes, which are generally protected by horny hoofs. Each stride is lengthened by the length of the hand or foot plus the toe bones which are often greatly elongated. Many long-distance runners such as horses and antelopes are unguligrade.

uric acid   A waste product formed from the breakdown of AMINO ACIDS. It is produced by animals such as insects, snails, reptiles and birds which need to conserve water rather than lose it in excreta.

urine   Liquid waste produced by the kidneys (and stored in the bladder) which contains mainly the breakdown products of AMINO ACIDS.

uterus   The muscular organ in female mammals and some other vertebrates in which the EMBRYOS develop.

V

vector   A blood-feeding insect, such as a mosquito, which transmits parasitic organisms, for which it is the intermediate HOST, to another host in the PARASITE's life cycle.

veliger   The LARVA of gastropods and some other mollusks, which develops from the TROCHOPHORE. The CILIATED region is drawn out into an enlongated velum, which supports the animal like water skis. During the veliger state TORSION takes place.

viviparous   Describes the development of an EMBRYO inside and nourished by the mother, until it is born, when it is a smaller version of the adult.

Z

zooplankton   The animal members of the PLANKTON. Many are LARVAE but some, such as jellyfish, include the largest members of the plankton.

zygote   The fertilized ovum before it begins further development.

INDEX

In this index, italic numerals (e. g. *70*) refer to illustrations or their captions. The symbol * indicates a topic that has an entry in the Glossary (pp. 146—152).

CREDITS

Michael Abbey/Okapia 15; Kurt Amsler/Seaphot 62; Heather Angel 12, 19, 20, 25, 26, 35, 37, 40, 41, 43, 44, 50, 53, 57, 60, 64, 71, 72, 73, 78, 88, 91, 96, 98, 114, 115, 116, 143; Anthony Baventock/Seaphot 39; James Bell/Science Photo Library 9; S. C. Bisserot FRPS/Nature Photographers Ltd 103, 105, 113; Frank V. Blackburn/Nature Photographers Ltd 18; P. Boston/Natural Science Photos 109; Brinsley Burbidge/Nature Photographers Ltd 142; N. A. Callow/Nature Photographers Ltd 59; Peter D. Capen/Seaphot 136; Dick Clarke/Seaphot 62; Colour Library International 43, 89, 90; Conor Craig/Seaphot 68; Richard Crane/Nature Photographers Ltd 135; Bill Curtsinger/Okapia 125; Martin Dohrn/Science Photo Library 55, 72; Georgette Douwma/Seaphot 67; Friskney Essex/Natural Science Photos 140; Douglas Faulkner/Okapia 127; Geoscience Features 33; Finnish Tourist Office 95; Ron and Christine Foord 51; Gower Medical Publishing 30, 53, 70, 95; Dr Steve Gull/Science Photo Library 63; Eric Gravé/Science Photo Library 17, 20, 32, 36; James Hancock/Nature Photographers Ltd 126, 135; Margaret Hayman/Seaphot 57; A. Hayward/Natural Science Photos 74, 80; Jan Hirsch/Science Photo Library 24; J. Hobday/Natural Science Photos 101, 102; David Hosking 25; James Hudnall/Seaphot 117; Alan Hutchinson Library 123; Masao Kawai/Orion Press/ Bruce Coleman Ltd 108; Edgar T. Jones/Aquila Photographics 91; Nicholas Law 11, 133; Hugo van Lawick/Nature Photographers Ltd 119; J. Lawton Roberts/Aquila Photographics 90; Michael Leach/Nature Photographers Ltd 142; Ken Lucas/Seaphot 23, 31, 137; John and Gillian Lythgoe/Seaphot 122; R. L. Matthews/Seaphot 141; C. Mattison/Natural Science Photos 77; L. C. Marigo/Bruce Coleman Ltd 106; Nature Photographers Ltd 47; Peter Newark's Western America 134; Okapia 58, 97; W. S. Patou/Nature Photographers Ltd 139; J. M. Pearson/Biophotos 19, 132; Howard Platt/Seaphot 94; K. R. Porter/Science Photo Library 15; Jurge Provenza/Seaphot 106; W. J. von Puttkamer/Alan Hutchinson Library 100; Ian Redmond/Seaphot 59, 110; Brian Rogers/Biofotos 89; Rod Salm/Seaphot 76; Scala III; Peter Scoones/Seaphot 68, 69; Jonathan Scott/Seaphot 22, 27, 28, 29; Seaphot 44, 46, 47, 51; David Sewell/Nature Photographers Ltd 139; Silvestris/Meyers 120; Silvestris/Wothe 121; Dr Nigel Smith/Alan Hutchinson Library 100, 131; Sinclair Stammers/Science Photo Library 46; Tony Stone Associates 6, 8, 10, 34, 61, 66, 71, 79, 81, 92, 93, 96, 105, 119, 121, 128, 129, 131, 133, 141, 143; Soames Summerhays/Biofotos 12, 124, 139, 142; Herwarth Voigtmann/Seaphot 65; John Walsh/Science Photo Library 13, 38, 49; P. H. and S. L. Ward/Natural Science Photos 77, 115, 127; Gary Weber/Aquila Photographics 75; Curtis Williams/Natural Science Photos 87; Robin Williams/Gower Medical Publishing 95; York Museum 135.

Cover photo—Russ Kinne, Photo Researchers